Roadmap
to the MCAS:
Grade 8 Math

by Athlene Whyte-Smith

Random House, Inc.
New York

www.randomhouse.com/princetonreview

This workbook was written by The Princeton Review, one of the nation's leaders in test preparation. The Princeton Review helps millions of students every year prepare for standardized assessments of all kinds. The Princeton Review offers the best way to help students excel on standardized tests.

The Princeton Review is not affiliated with Princeton University or Educational Testing Service.

Princeton Review, L.L.C.
160 Varick Street, 12th floor
New York, NY 10013

E-mail: textbook@review.com

Published in the United States by Random House, Inc., New York.

ISBN 0-375-76368-6

Editor: Linda Fan
Development Editor: Scott Bridi
Director of Production: Iam Williams
Design Director: Tina McMaster
Art Director: Neil McMahon
Production Editor: Lisbeth Dyer
Production Coordinator: Alexandra Morrill

Manufactured in the United States of America.

9 8 7 6 5 4 3 2 1

First Edition

Acknowledgments

I would like to thank my husband and children for their patience, understanding, and love as I toiled night after night (and day after day).

I would also like to thank the editorial and production staff at The Princeton Review.

CONTENTS

INTRODUCTION

INTRODUCTION FOR PARENTS AND TEACHERS

ABOUT THE PRINCETON REVIEW

The Princeton Review is one of the nation's leaders in test preparation. We prepare more than two million students every year with our courses, books, on-line services, and software programs. In addition to helping students in Massachusetts with the Grade 8 MCAS Math test, we make study guides for the Grade 10 MCAS Math test, and the Grade 7 and Grade 10 MCAS English Language Arts tests. We also coach students around the country on their statewide standardized tests and on college entrance exams such as the SAT-I, SAT-II, PSAT, and ACT. Our strategies and techniques are unique and, most importantly, successful. Our goal is to reinforce skills that students have been taught in the classroom and show them how to apply those skills to the specific format and structure of the Grade 8 MCAS Math test.

ABOUT THIS BOOK

Roadmap to the MCAS: Grade 8 Math contains three main parts: an introduction, lessons in which students practice individual math skills, and practice tests.

This introduction explains the particulars of the Grade 8 MCAS Math test, including vital information such as the tools students are allowed to use while taking the test and the amount of time they will have to finish the test.

The lessons in this book are called "miles" to go with the roadmap theme. Each mile focuses on a specific math skill that is included in the Massachusetts eighth-grade math curriculum. The thirty-nine miles in this book include skills that will be tested on the Grade 8 MCAS Math test. The first three miles provide techniques students could use to answer the different types of questions that will appear on the test: multiple choice, short answer, and open response. The answer key for the miles begins on page 143.

This book also contains two full-length practice tests that are modeled on the Grade 8 MCAS Math test. Instructions for taking the practice tests appear on page 160. By administering the practice tests in test-like conditions, you will help students become familiar with the testing situation they will experience when they take the Grade 8 MCAS Math test. Reviewing students' performance on the practice tests will help you assess which skills they need more practice with before they take the MCAS test. Answer keys and explanations for the correct answers to the questions on the practice tests begin on page 237.

Before students begin working through this book, look through the content of the miles. We realize that every student, and every class, has different strengths and weaknesses. There will be no harm done if students work on the miles out of sequence.

ABOUT THE MASSACHUSETTS COMPREHENSIVE ASSESSMENT SYSTEM (MCAS)

All public school students in Massachusetts must take MCAS tests. In grade eight, students take tests in mathematics, science, and history. Results of the tests will be used in several ways. Parents and educators may use the test scores to monitor students' progress and to identify weaknesses in curriculum and instruction. In grade ten, students take MCAS tests in mathematics and English language arts. Passing the grade ten MCAS tests is required for a high school diploma. If necessary, students will be given multiple opportunities to pass the grade ten tests.

For additional information about testing and curricula, visit the Massachusetts Department of Education's Web site: www.doe.mass.edu. The Web address for the MCAS home page is www.doe.mass.edu/mcas.

Information about MCAS administration for students with disabilities and students with limited English proficiency may be found on the Internet at www.doe.mass.edu/mcas/part_req.html.

ABOUT THE GRADE 8 MCAS MATH TEST

Students will take the Grade 8 MCAS Math test in May. The testing dates vary annually, so you should check your school's schedule before students take the test.

The test includes two sessions and forty-eight total questions. Students will have to answer thirty-six multiple-choice questions, six short-answer questions, and six open-response questions on the test. Students will have approximately sixty minutes to answer the questions in each of the two sessions on the test. Extra time is available; check the MCAS home page or ask your student's school.

Session 1 includes multiple-choice, short-answer, and open-response questions. While answering the questions in Session 1, students are not allowed to use calculators.

Session 2 includes multiple-choice and open-response questions, but no short-answer questions. While answering the questions in Session 2, students are allowed to use a four-function calculator with a square root key. Calculators may be provided to students who do not have them while they take the test.

Students will be provided with a mathematics reference sheet on the test. The reference sheet includes formulas, conversions, and a ruler. A reference sheet similar to the one that students will use on the Grade 8 MCAS Math test is on page 7.

For both sessions of the test, students are allowed to write in their test books. This is an advantage to students because it allows them to write out their calculations right next to the problems they are solving.

INTRODUCTION FOR STUDENTS

ABOUT THIS BOOK

This book is designed to help you improve your score in your eighth-grade math class and on the Grade 8 MCAS Math test. In this book, you will review and practice the skills that you should know for your exam and for your eighth-grade math class. So you can learn to do well on the test and in your class at the same time.

There are three main parts of this book: this introduction, the miles, and the practice tests. In the miles, you will review and practice specific math skills. The practice tests give you an opportunity to take tests that are similar to the actual Grade 8 MCAS Math test.

ABOUT THE GRADE 8 MCAS MATH TEST

All eighth-graders in Massachusetts public schools have to take the Grade 8 MCAS Math test. The purpose of the test is to show you and your teachers and parents what you know. The results of the test will show which math skills you know well and which ones you might need a little help on.

You will take the test sometime in May while you are in the eighth grade. Ask your teacher for the exact testing dates if you don't know them. The test includes forty-eight total questions. You will have to answer thirty-six multiple-choice questions, six short-answer questions, and six open-response questions on the test. You will have sixty minutes to answer the questions in each of the two sessions on the test. That means you'll have two hours to answer all forty-eight questions on the test.

The test is separated into two sessions. Session 1 includes multiple-choice, short-answer, and open-response questions. While answering the questions in Session 1, you are **not** allowed to use a calculator. Session 2 includes multiple-choice and open-response questions, but no short-answer questions. While answering the questions in Session 2, you are allowed to use a four-function calculator with a square root key. You should bring a calculator with you when you take the test.

You will be provided with a mathematics reference sheet to use while you take the test. The reference sheet includes formulas, conversions, and a ruler. There is a reference sheet similar to the one that you will use on the Grade 8 MCAS Math test on page 7.

For both sessions of the test, you are allowed to write in your test booklets. This is an advantage because it allows you to write out your calculations right next to the problems you are solving.

PREPARING FOR THE GRADE 8 MCAS MATH TEST

Here is a list of things you can do to prepare for the Grade 8 MCAS Math test.

- **Ask questions.** If you are confused after you finish working on a mile (or even while answering just one question), ask a parent or teacher for help. Asking questions is the best way to make sure that you understand what you have to do in order to do well on the test.

- **Practice.** Math is all around you. Every time you go to the grocery store, make dinner, or play sports, math comes along for the ride. Use real-life opportunities to practice the math you know. Add up the prices of the groceries in your cart while you're waiting on line. See if your addition matches what comes up on the register. If you play softball or baseball, keep track of your batting average. Every day there will be more math to do, and the great thing is that it's the same type of math that will be on the Grade 8 MCAS Math test.

- **Read.** Read everything you can. Read the newspaper, magazines, books, plays, poems, comics, and even the back of your cereal box. The more you read, the better you will become at it. And the better you read, the more likely you will be to do well on the Grade 8 MCAS Math test. Many of the questions on the test will be word problems. If you can't read and understand the problems, you won't have a chance to show all the math you know.

- **Eat well and get a good night's sleep.** Your body doesn't work well when you don't eat good food and get enough sleep. Neither does your brain. On the night before the test, make sure to go to bed at your normal time and get plenty of sleep. You should also eat a healthy breakfast on the morning of the testing day. It is important to be awake and alert while you take the Grade 8 MCAS Math test, or any other test for that matter.

This is just the beginning of the road. There are great things to learn ahead. So buckle your seat belt and get ready to travel the first mile to grade eight math excellence.

MATHEMATICS REFERENCE SHEET

The mathematics reference sheet below is very similar to the one you will see on the Grade 8 MCAS Math test. You will use the reference sheet to answer some questions on the test. You may use the sheet below to answer some questions in the miles and practice tests in this book.

PERIMETER FORMULAS

square.........$P = 4s$

rectangle.........$P = 2b + 2h$

triangle.........$P = a + b + c$

Pythagorean theorem

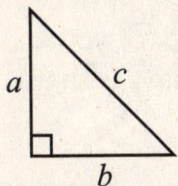

$$a^2 + b^2 = c^2$$

CIRCUMFERENCE FORMULAS

circle.........$C = 2\pi r$

OR

$C = \pi d$

CONVERSIONS

1 mile = 5,280 feet

1 square mile = 640 acres

AREA FORMULAS

square.........$A = s^2$

rectangle.........$A = bh$

OR

$A = lw$

triangle.........$A = \frac{1}{2}bh$

circle.........$A = \pi r^2$

trapezoid.........$A = \frac{1}{2}h(b_1 + b_2)$

VOLUME FORMULAS

rectangular prism......$V = Bh$
(B = area of base)

cone....................$V = \frac{1}{3}\pi r^2 h$

cylinder.................$V = \pi r^2 h$

cube....................$V = s^3$
(s = length of an edge)

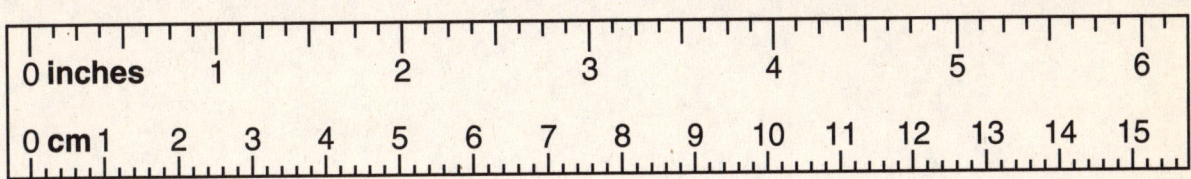

MILES

MILE 1: ANSWERING MULTIPLE-CHOICE QUESTIONS

The Grade 8 MCAS Math test consists of forty-eight questions, thirty-six of which are multiple-choice. Each multiple-choice question is followed by four possible answer choices. You have to pick the best answer choice for each question.

You should approach multiple-choice questions like any other math problems. First, carefully read the question and figure out what you need to do in order to answer it. You may have to perform calculations to figure out the answer. What makes multiple-choice questions different is that they come with four answer choices, one of which is the correct answer. Once you figure out your answer, just fill in the bubble that goes with it.

But what happens if it's not as easy as that? That's where the Process of Elimination comes in.

PROCESS OF ELIMINATION (POE)

The fact that most questions on the Grade 8 Math MCAS test are multiple-choice questions is a good thing. Why? Because the correct answer choice is there. If you solve a question but can't find your answer among the choices, then you've made an error. Go back and check your math. But even if you're not sure which answer choice is correct you can often use Process of Elimination (POE) to find it. How does this work? First, look for answer choices that you *know* are wrong. Then, get rid of those answer choices by crossing them off with your pencil. Look at an example to see how this works.

▶ **What is the capital of Burundi?**

_____

Not sure? Without any answer choices, you would have to guess (and you would probably be wrong). But with multiple-choice questions, you just have to pick the answer choice that you think is best. If you look closely at the answer choices, they often give you valuable information. Now look at the same example question with answer choices.

▶ **What is the capital of Burundi?**

A. New York City

B. Bujumbura

C. London

D. Boston

Can you guess now? Boston is the capital of Massachusetts, so D is clearly not right. London is the capital of England, so C can't be right, either. And New York City, while not the capital of New York, probably isn't the capital of Burundi, either, so you can eliminate A as well. Even if you've never heard of the country of Burundi, you can probably recognize that A, C, and D are *not* its capital. If your geography teacher stopped you in the hall and gave you a pop quiz, you'd be stumped. But on a test with a multiple-choice format, you can get this question right! Singling out Bujumbura as the correct answer choice is an example of using POE.

POE works especially well with math questions. Look at a question that could be difficult to solve without using POE.

▶ **What is the value of 0.4 × 0.5 as a fraction?**

A. $\frac{1}{100}$

B. $\frac{1}{5}$

C. $\frac{9}{20}$

D. $\frac{8}{3}$

You could spend a lot of time calculating the answer for this question. Instead, look at the answer choices. Notice that they are fairly different from each other. When this is the case, you can attack the answer choices to find the correct answer.

First make a quick estimate. 0.4 is close to $\frac{1}{2}$ and 0.5 is equal to $\frac{1}{2}$. Multiply those fractions together to get an approximation of the answer, $\frac{1}{2} \times \frac{1}{2} = \frac{1}{4}$. Your answer should be about $\frac{1}{4}$. Choice A is much smaller than $\frac{1}{4}$; eliminate it. Choice B is about right. Keep it and go on to the next answer choice. Choice C is about $\frac{1}{2}$. It's too large; get rid of it. Finally, look at choice D. It is greater than 1. This is not the correct answer. The only possible answer is B, which you can arrive at without performing a lot of calculations! The key to solving some math questions on the multiple-choice test is attacking the answer choices.

What if you can't get rid all of the wrong answer choices? That's okay. You will have a better chance of picking the correct answer choice if you can get rid of even one wrong answer choice. On the MCAS, you always start with four answer choices. If you have no idea how to answer a question, you have only a one in four or 25% chance of guessing the correct one. If you get rid of even one or two answer choices, your odds of picking the correct answer improve to either 33% or 50%.

You've now seen how powerful POE can be. Although you probably won't be able to eliminate all three wrong answers for *every* question, crossing out even one or two incorrect answer choices can help to improve your score dramatically on any multiple-choice test that you take.

POE can have a significant impact on your test score. You should combine your mathematical skills with POE to maximize your score on every multiple-choice test.

MILE 2: ANSWERING SHORT-ANSWER QUESTIONS

In addition to multiple-choice questions, there are short-answer questions on the MCAS. Unlike multiple-choice questions, where the correct answer is listed as one of the answer choices, short-answer questions will require you to write your own response.

You should approach these questions slightly differently from the multiple-choice questions. Because there are no answer choices, you will not be able to use Process of Elimination. Solve the question and take extra time to check your answer. Once you have an answer, be sure to write it in the designated area on the answer booklet.

Look at an example of a short-answer question.

▶ **T-shirts are on sale for 20% off the regular price. If the regular price of a T-shirt is $15, what is the sale price?**

Step 1 First, read carefully and understand what the question is asking. This question asks you to find the sale price of a T-shirt given the regular price and the discount. You may write on your test booklet, so circle or underline the key information: *20% off* and *$15.*

Step 2 You may solve this problem in a number of ways. One way is to find the amount of the discount and subtract it from the regular price. The discount is 20% of the regular price of $15, or 20% × $15. Change the percent to a decimal before multiplying: 0.20 × $15 = $3. This is the amount taken off the regular price. Now subtract $3 from $15 to find the sale price: $15 − $3 = $12. The sale price is $12.

Step 3 Check your work. You may ask yourself if this solution makes sense. The sale price is lower than the original price, but not too low. This is a reasonable answer. Remember that the discount is only 20%. If you had written the discounted amount, $3, as the answer, you should have realized that this is much too low.

Step 4 Write your answer in the answer box above. On the test, write your answer in the correct box in the answer booklet. Write only your answer in the answer box. Do all your work in the test booklet.

Look at another example of a short-answer question.

▶ **Mindy is paving her walkway with concrete. The rectangular walkway is 6 yards long and 2 yards wide. She will use a concrete that costs $1.25 per square foot. How much will it cost to have the walkway paved?**

Step 1 First, read the question carefully. You are asked to find how much it will cost Mindy to pave her walkway. Circle or box the key pieces of information. You may circle *6 yards long, 2 yards wide,* and *$1.25 per square foot.*

Step 2 Plan how you will answer this question. You need to find the area to be paved and then multiply it by the cost. Notice that the cost of the concrete is given in square feet, and that the units for the walkway dimensions are given in yards. Make a decision: find the area of the walkway in square feet or find the price of the concrete per square yard. Either method will get you the same answer; it's up to you.

Find the area of the walkway in square feet. To change yards to feet, remember that there are 3 feet in a yard. Multiply each dimension by 3. The walkway is $6 \times 3 = 18$ feet long and $2 \times 3 = 6$ feet wide. The area of the walkway is length times width, 18 feet $\times$ 6 feet, or 108 square feet. (Area is covered in Mile 34.)

Once you know the area to be paved, multiply the area by the cost per square foot. 108 feet $\times$ $1.25 = $135. The cost to have the walkway paved is $135.

Step 3 Check your work and see if it makes sense. If you have time, you may want to solve this problem another way and make sure you get the same answer. Otherwise, estimate to see if your answer is in the ballpark. The area of the walkway is about 20 feet by 6 feet, or 120 square feet. The price of the concrete is about $1 per square foot, so the answer should be about $120. Your answer is reasonable.

Step 4 Write the answer in the answer box above. On the test, remember to write your answer in the correct answer box on your answer booklet. Write only your answers in the answer box. Do all your work in the test booklet.

For short-answer questions, you will come up with your own answers, so take a few seconds to make sure your answer makes sense! Also, you won't get credit if you don't write your answer in the correct space in the answer booklet. Don't go on to the next question until you transfer your answer to the answer booklet.

MILE 3: ANSWERING OPEN-RESPONSE QUESTIONS

Questions on the MCAS will ask you to solve open-response questions. Open-response questions are multiple-step problems that require you to generate your own written responses. You may be asked to explain your answer in one to two paragraphs or to create a table, chart, graph, diagram, or illustration.

Open-response questions do not provide answer choices. They are different from the multiple-choice questions, where you choose an answer from the given answer choices. Unlike the short-answer questions, there are multiple parts to each open-response question. You will need to write an answer to each part of the question in the designated area on the answer booklet.

The open-response questions are scored using a 4-point scale or rubric. It is important that you show all your work when answering open-response questions. You can receive points for understanding the math concepts in the question even if you don't answer all parts of the question correctly. So if you performed some of the calculations correctly and demonstrate an understanding of the question, you can receive points. The reverse is also true. You can get the final answer correct and only receive partial credit if you do not show all the necessary work.

Look at an example of an open-response question and its solution.

▶ **Ali, Eliot, Paul, and Tina are each helping to decorate a different float for the All-City Parade. The order in which the floats will appear is determined randomly by a drawing. Each float has an equal chance of going first, second, third, or fourth.**

a. **What is the probability that Eliot's or Tina's float will go first?**

b. **Show all the possible orders in which the floats can appear if Tina's float goes first. Use a tree diagram, chart, or list. You may use the first letter of each student's name.**

c. **What is the probability that when Tina's float appears first, Eliot's float will appear second and Ali's float will appear third? Explain your answer.**

For open-response questions, you will need to answer each part and show all your work in the answer box provided. In the Miles section of this book, you may use a separate sheet of paper to show your work. See how you could find the answer to each part of this question.

Part a. You are asked for a probability, the ratio of the number of favorable outcomes to the total number of possible outcomes. The outcome that we are interested in is whether Eliot's or Tina's float will go first. There are four floats in total, so there is a one in four chance that Eliot's float will go first and a one in four chance that Tina's float will go first. Add to find the probability that one of those two floats will go first in the parade: $\frac{1}{4} + \frac{1}{4} = \frac{2}{4}$, or $\frac{1}{2}$.

In the answer box provided in your answer booklet, you should write

    a. *Probability that Eliot's float will go first* $= \frac{1}{4}$

       Probability that Tina's float will go first $= \frac{1}{4}$

       The probability that Eliot's or Tina's float will go first is $\frac{1}{4} + \frac{1}{4} = \frac{2}{4}$ *or* $\frac{1}{2}$.

Part b. You can make an organized list to show the possible orders of the floats if Tina's float goes first. Because Tina will always occupy the first space, write *T* as the first letter and then find all the arrangements of the other letters.

In the answer box provided in your answer booklet, you should write

    b. *If Tina's float goes first, all the possible outcomes include*

       TAEP   TEAP   TPAE

       TAPE   TEPA   TPEA

Part c. This part asks for the probability that a particular order of floats will appear in the parade. Because the question specifies that Tina's float goes first, use the organized list you made for part b to answer this question. You are looking for the number of times you will see Tina, Eliot, and Ali in that order. Notice that this order only occurs 1 time out of the 6 possible outcomes when Tina goes first. The probability that the floats will occur in the given order is $\frac{1}{6}$.

In the answer box provided in your answer booklet, you should write

    The probability that the floats will occur in the given order is $\frac{1}{6}$. *There are 6 possible float arrangements when Tina goes first. Of those arrangements, there is only one where Eliot's float appears second and Ali's float appears third.*

Again, show all your work and write your explanations clearly when tackling open-response questions. Write what you know even if you cannot answer the entire question. You may earn partial credit.

MILE 4: REAL NUMBERS

Whole numbers are the counting numbers and zero. They form the set {0, 1, 2, 3, . . .}.

Integers are positive and negative whole numbers and zero. They form the set {. . . –2, –1, 0, 1, 2, . . .}.

Rational numbers are all the numbers that can be expressed as a fraction, $\frac{a}{b}$, when a and b are integers and $b \neq 0$. All whole numbers and integers are rational numbers because they can be represented as fractions. For example, $\frac{2}{1} = 2$ and $-\frac{2}{1} = -2$. Rational numbers also include terminating or repeating decimals because they can be represented by fractions. (A line, or bar, over a number means that it repeats.)

Example 1: $0.3333333\overline{3}$

This decimal ends in a repeating pattern. It is a rational number because it can be written as $\frac{1}{3}$.

Irrational numbers are numbers that cannot be written as $\frac{a}{b}$, when a and b are integers and $b \neq 0$. Irrational numbers are decimal numbers that *never* terminate or repeat.

Example 2: $0.353353335333353333335 \ldots$

The decimal above continues in a pattern that never ends or repeats. It is an irrational number. Even though the example above ends in a predictable pattern—the number of 3s increases—it is not rational because the pattern doesn't repeat exactly.

Real numbers are the set of rational numbers together with the set of irrational numbers.

Look at the chart below for more examples of real numbers.

$0.22322242223222225\ldots$	Irrational, because it never terminates or repeats
$\frac{2}{5}$	Rational, because it can be expressed as $\frac{a}{b}$, where $a = 2$ and $b = 5$, and because it is a terminating decimal 0.4
π or $3.141592654\ldots$	Irrational, because it never terminates or repeats
$\sqrt{49}$	Rational, because it is equal to whole number 7, which can be expressed as $\frac{7}{1}$
$\sqrt{5}$	Irrational, because it never terminates or repeats: $\sqrt{5} = 2.2360679\ldots$
$0.\overline{3}$	Rational, because it is a repeating decimal that can be expressed as $\frac{1}{3}$

Directions: You will need a calculator for this activity. Solve each of the problems below using your knowledge of real numbers. Each answer corresponds to a letter on the number line below. As you solve each problem, place the letter in the corresponding box at the bottom of the page. If you match the numbers and letters correctly, you will figure out the answer to the riddle at the bottom of the page.

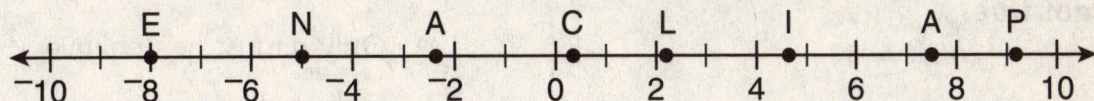

1. What is the approximate value

   of $^-\sqrt{6}$? _____

   Is $^-\sqrt{6}$ a rational number? _____

2. What is the approximate value

   of $\frac{64}{7}$? _____

   Is $\frac{64}{7}$ a rational number? _____

3. What is the value of $^-\sqrt{64}$?

   _____

   Is $^-\sqrt{64}$ a rational number? _____

4. What is the approximate value

   of $\sqrt{5}$? _____

   Is $\sqrt{5}$ a rational number? _____

5. What is the value of $\sqrt{\frac{24}{6}} + \frac{8}{3}$?

   _____

   Is $\sqrt{\frac{24}{6}} + \frac{8}{3}$ a rational number?

   _____

6. What is the approximate value

   of $\frac{1}{3}$? _____

   Is $\frac{1}{3}$ a rational number? _____

7. What is the approximate value

   of $\sqrt{56}$? _____

   Is $\sqrt{56}$ a rational number? _____

8. What is the approximate value of

   $\sqrt{56} \times \frac{^-\sqrt{75}}{13}$? _____

   Is $\sqrt{56} \times \frac{^-\sqrt{75}}{13}$ a rational

   number? _____

What kind of can never needs a can opener?

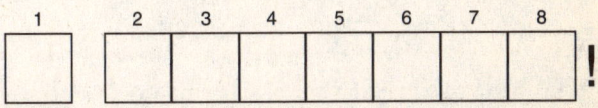

Directions: Read each question carefully. Circle your answer choice for each question.

1 For the equation below, which of the following statements is **not true?**

$$(2ab)^2 = 37$$

A. Both a and b can be positive.

B. Both a and b can be negative.

C. If a is negative, then b must be positive.

D. If a is negative, then b can be positive or negative.

2 In the equation $\frac{n}{4} = -\frac{1}{2}$, which of the following **best** describes n?

A. a negative integer

B. a positive integer

C. a whole number

D. an irrational number

3 In the equation $x^2 = 26$, which of the following **best** describes x?

A. an integer

B. a whole number

C. a rational number

D. an irrational number

4 If $\frac{p}{q} = -2$, which of the following is a true statement?

A. Only p must be negative.

B. Only q must be negative.

C. Either p or q must be negative.

D. Both p and q must be negative.

5 In which of the following equations could the variables both be negative numbers?

A. $2st = -14$

B. $-2st = -14$

C. $2s = -14t$

D. $-2s = 14t$

When trying to figure out whether a statement is true, use real number examples to check each answer choice. If you can make the statement false, cross off that answer choice and move on to the next one.

6 Which of the following always has a product of 1?

  A. p and $-p$

  B. p and $-\frac{1}{p}$

  C. p and $\frac{1}{p}$

  D. p and p

7 What must always be the product of x and $-\frac{1}{x}$?

  A. -1

  B. 0

  C. 1

  D. x^2

8 The variable n cannot be 0 in which of the following expressions?

  A. $\frac{n}{5}$

  B. $5n$

  C. $n - 5$

  D. $\frac{5}{n}$

9 If a is a positive number, which of the following does not represent a real number expression?

  A. $-\sqrt{a}$

  B. $\sqrt{-a}$

  C. $\frac{a}{b}$

  D. $\sqrt{\frac{a}{b}}$

10 The product of two nonzero numbers will result in which of the following?

  A. only a positive number

  B. only a negative number

  C. only a rational number

  D. only a real number

MILE 5: PROPERTIES OF NUMBERS

In this mile you will practice using the associative, commutative, and distributive properties in different mathematical expressions. You will also identify inverses and identities in numeric and algebraic expressions. Remember that an algebraic expression is a combination of variables (letters in italics that can represent any number), numbers, and a minimum of one operation, such as multiplication or addition. A numeric expression is a combination of only numbers and at least one operation.

Associative property: For any numbers a, b, and c, you will get the same result no matter how you *group* the numbers when adding or multiplying.

$$(a + b) + c = a + (b + c) \quad \text{and} \quad (a \bullet b) \bullet c = a \bullet (b \bullet c)$$
$$(3 + 4) + 8 = 3 + (4 + 8) \quad \text{and} \quad (3 \bullet 4) \bullet 8 = 3 \bullet (4 \bullet 8)$$

Commutative property: For any numbers a and b, you will get the same result no matter how you *order* the numbers when adding or multiplying.

$$a + b = b + a \quad \text{and} \quad a \bullet b = b \bullet a$$
$$7 + 4 = 4 + 7 \quad \text{and} \quad 7 \bullet 4 = 4 \bullet 7$$

Distributive property: For any numbers a, b, and c, where one number is multiplied by a sum of two addends, you will get the same result when you multiply that number by each of the two addends and add the products.

$$a \bullet (b + c) = (a \bullet b) + (a \bullet c)$$
$$2 \bullet (5 + 6) = (2 \bullet 5) + (2 \bullet 6)$$

Identity property: If you add 0 to a number or multiply a number by 1, the result is the same number. The identity or value of the number stays the same.

$$a + 0 = a \quad \text{and} \quad a \times 1 = a$$
$$3 + 0 = 3 \quad \text{and} \quad 3 \times 1 = 3$$

Inverse property of addition: The sum of an integer and its additive inverse is equal to 0. Remember that the additive inverse of a number is the same distance away from zero, but in the opposite direction. So the opposite of –10 is 10.

$$b + (-b) = 0 \qquad 3 + (-3) = 0$$

Inverse property of multiplication: The product of a number and its multiplicative inverse is equal to 1. Multiplicative inverses are also called reciprocals.

$$\frac{4}{5} \bullet \frac{5}{4} = 1 \qquad \frac{a}{b} \bullet \frac{b}{a} = 1, \text{ where } a \neq 0 \text{ and } b \neq 0$$

Multiplicative property of zero: Any number multiplied by 0 will equal 0.

$$a \bullet 0 = 0 \qquad \text{and} \qquad 6 \bullet 0 = 0$$

Directions: Match the mathematical expression in the left column with the property it demonstrates in the right column. Each property should match an expression only once. If you match the items in the two columns correctly, you will spell out the answer to the question at the bottom of the page.

1. _____ $1.3 + 5 = 5 + 1.3$ R) Associative property of addition

2. _____ $-\frac{3}{4} \times -\frac{4}{3} = 1$ E) Identity property of multiplication

3. _____ $3(5 + 6) = 15 + 18$ O) Identity property of addition

4. _____ $8 \times 2 = 2 \times 8$ T) Distributive property

5. _____ $6 \times 1 = 6$ M) Commutative property of addition

6. _____ $(4 + 6) + 2 = 4 + (6 + 2)$ R) Associative property of multiplication

7. _____ $6 + -6 = 0$ H) Inverse property of addition

8. _____ $14 + 0 = 14$ A) Inverse property of multiplication

9. _____ $(2 \times 4)3 = 2(4 \times 3)$ N) Multiplicative property of zero

10. _____ $17 \times 0 = 0$ T) Commutative property of multiplication

According to some lists, this mountain is one of the Seven Natural Wonders of the World. It is shaped like a pyramid and is located in Italy and Switzerland. What is it called?

_____

Directions: Read each question carefully. Circle your answer choice for each question.

1 Which of the expressions shown below is the same as $4(y - 6)$?

A. $4y + 24$

B. $y^A - 6$

C. $4y - 6$

D. $4y - 24$

2 Which of the following expressions is true if $c + b = a$?

A. $a + c = b$

B. $b \cdot c = a$

C. $b + c = a$

D. $\frac{a}{c} = b$

3 Which mathematical property is shown by $-54 + 54 = 0$?

A. commutative property

B. inverse property

C. identity property

D. distributive property

4 Which of the mathematical expressions below shows an example of the associative property?

A. $(4 \times 5)9 = 4(5 \times 9)$

B. $4 \times 0 = 0$

C. $4 \times 1 = 4$

D. $11(8 - 3) = 11(8) - 11(3)$

5 Which of the following expressions is another way to write $a(bc)$?

A. $ab + ac$

B. $(ab)c$

C. $ab + c$

D. $(ab)(ac)$

Tip

Think about the meanings of the names of the properties to help you remember them. A number maintains its *identity* after it is multiplied by 1 or added to 0, so that's the **identity property**. The **associative property** applies to numbers that are *associated* with each other, or grouped together.

6 Linda wants to use the distributive property to simplify the problem shown below. Which of the following answer choices shows a proper application of the distributive property?

$$\frac{5}{8}(4 + 8) =$$

A. $\frac{5}{8} + 12$

B. $\frac{5}{8}(4) + 8$

C. $\frac{5}{8}(4) + 5$

D. $\frac{5}{8} - 12$

7 Which of the following expressions is an example of the distributive property?

A. $2 \bullet 4 = 4 \bullet 2$

B. $(54 + 43) + 2 = 54 + (43 + 2)$

C. $4(34 + 2) = 4(34) + 4(2)$

D. $3 \bullet 1 = 3$

8 Which algebraic expression is the same as $a(b + c)$?

A. $a + b + c$

B. $ab + ac$

C. $(ab)(ac)$

D. $ab + c$

9 Scott and Nick are working on their math homework without a calculator. Which property could they use simplify the problem below?

$$\frac{6}{7} \times \frac{7}{6} \times 8 \times 2 =$$

A. Identity property of addition

B. Identity property of multiplication

C. Commutative property of multiplication

D. Inverse property of multiplication

10 Which of the expressions below is an example of the inverse property?

A. $\frac{4}{6} \times \frac{6}{4}$

B. 6×-6

C. $6(4) = 4(6)$

D. $6(4 + 1) = 6(1 + 4)$

11 Using the multiplicative property of zero, what is the value of 125×0?

A. -125

B. 0

C. 1

D. 125

MILE 6: OPERATIONS

To solve most math problems, you will have to use one or more operations. The four primary **operations** are addition, subtraction, multiplication, and division. Some problems may be a straightforward calculation question where the operation signs are given. Some questions may be word problems where you will have to figure out which operation to use.

Fortunately, you will not be left in the dark when figuring out the operations for word problems. There will clues in the problem that can help you decide which operations to use. Some of these clue words or actions are listed below.

- Addition: *combine, total, sum*

- Subtraction: *more than, take away, difference*

- Multiplication: *combine equal groups, times, of*

- Division: *make equal groups*

Look at a couple of example problems.

Example 1: **Last week Omar read 144 pages. This week he read 167 pages. How many pages did he read in total in these two weeks?**

To solve this problem, think of the action that is taking place. You are finding the total number of pages Omar read in two weeks. When you see *total,* think addition: 144 + 167 = 311. Omar read 311 pages in two weeks.

Example 2: **Fiona's mother was preparing to purchase a used car. The car dealership offered her two plans for payment. She could pay the full amount of $5,800, or she could pay on an installment plan. The installment plan meant that Fiona's mother would have to pay $700 up front, and then pay $240 for each of the next 24 months. How much more would the car cost if Fiona's mother chose the installment plan?**

Again, to solve this problem, think of the actions that are taking place. Fiona's mother would be making a number of equal payments on the installment plan, so multiply: $240 × 24 = $5,760. Combine the total equal payments with the amount her mother initially paid to find the total amount of money she would pay on the installment plan: $5,760 + $700 = $6,460. Finally, find how much more this payment plan costs than paying the full amount by subtracting: $6,460 − $5,800 = $660. The car would cost $660 more if Fiona's mother paid using the installment plan.

Directions: Read each question carefully. Write your answer on another sheet of paper or circle the letter next to your answer choice.

1

Debra and five friends decided to buy pencils. Which statement correctly compares the cost of buying 6 individual pencils to buying a pack of 8 pencils?

A. It costs $2.40 more to buy a pack of 8 pencils.

B. It costs $2.40 more to buy 6 individual pencils.

C. It costs $2.85 more to buy a pack of 8 pencils.

D. It costs $2.85 more to buy 6 individual pencils.

2 Marco rides the bus to and from school Monday through Friday. He uses two tickets each day. Which statement correctly compares the cost of using individual tickets to using a 7-day pass for Marco to ride the bus to school?

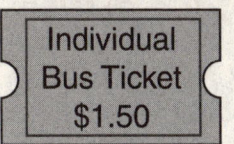 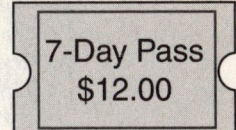

A. It costs $3.00 more to use the individual tickets.

B. It costs $10.50 more to use the individual tickets.

C. It costs $3.00 more to use the 7-day pass.

D. It costs $10.50 more to use the 7-day pass.

3 Karrie found the following coins in her backpack:

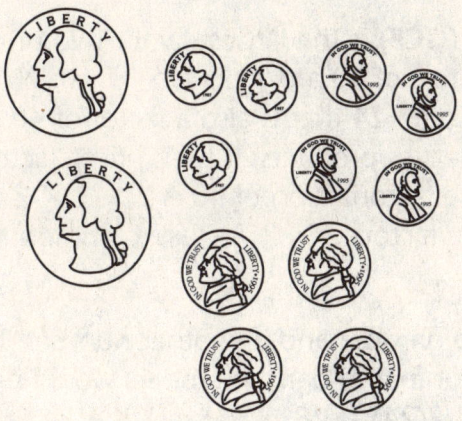

a. Karrie wanted to buy a pencil sharpener from the school store that costs $0.50. What combination of coins could she use? List all the possible combinations.

b. Can Karrie make 78 cents using exactly 6 coins from her backpack? Explain your answer.

c. Karrie wants to spend exactly 42 cents. What is the least number of coins she can use? Explain your answer.

MILE 7: MULTIPLES AND FACTORS

Suppose you wanted to find the different ways that 24 members of a school band could create an even marching formation. Factors will help you find the possible combinations.

A **factor** is a number that is multiplied to get a product. If you consider the total number of band members, 24, as the product, you can see that the band members can march in 1 row of 24 members, 2 rows of 12 members, 3 rows of 8 members, and so on. Each number that evenly goes into 24 is a factor. The number 24 has a total of 8 factors: 1, 2, 3, 4, 6, 8, 12, and 24.

A **prime number** is a number greater than 1 with only two factors—itself and 1. For example, the number 3 is a prime number because 1 and 3 are its only factors. Look at the factors of 24; only 2 and 3 are prime numbers.

The **greatest common factor (GCF)** is the largest factor two or more numbers have in common. For example, the factors of 40 are 1, 2, 4, 5, 8, 10, 20, and 40. The GCF of 24 and 40 is 8. This is the greatest factor of 24 that is also a factor of 40. You can also figure this out by rewriting the original numbers as products of their prime factors. The prime factorization of 24 is $2 \times 2 \times 2 \times 3$. The prime factorization of 40 is $2 \times 2 \times 2 \times 5$. Now look for factors common to both numbers. The factors $2 \times 2 \times 2$ are common to both numbers, so 8 is the GCF of 24 and 40.

Multiples are the products of a number and any other number. Think of other schools having 24-member bands. The total number of band members would be the multiples. The first few multiples of 24 are $1(24) = 24$, $2(24) = 48$, $3(24) = 72$, $4(24) = 96$, and $5(24) = 120$. Notice that you can have an unlimited number of multiples because you can always multiply one number by another number.

The **least common multiple (LCM)** is the smallest number that is divisible by two or more given numbers. For example, the multiples of 40 are the set {40, 80, 120, 160, . . .}. The LCM of 24 and 40 is 120. This is the smallest multiple that 24 and 40 have in common. You can also find the LCM of these numbers by looking at their prime factors again. Take the factors the numbers have in common ($2 \times 2 \times 2$) and the different factors (3 and 5), and multiply them for the final answer of $2 \times 2 \times 2 \times 3 \times 5$, or 120.

Directions: Read each question carefully. Write your answer on a separate sheet of paper or circle the letter next to your answer choice.

1 Richard has to arrange 100 bottles of soda into a display for the grocery store. One possible display he is considering is 10 rows with 10 bottles in each because he knows that 10 × 10 = 100. What is another design that would work?

   A. 2 rows of 40 bottles each

   B. 5 rows of 20 bottles each

   C. 10 rows of 8 bottles each

   D. 15 rows of 6 bottles each

2 Three planets are orbiting their star. The amount of time it takes the stars to make one revolution are 8 Earth years, 9 Earth years, and 12 Earth years. If the three planets are currently aligned in a straight line with their star, how many Earth years must pass before this alignment occurs again?

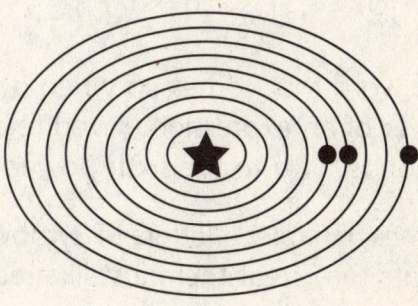

   A. 17

   B. 29

   C. 72

   D. 144

3 Mina's favorite number is between 100 and 200. It is a product of two prime numbers. One of its factors is 19.

   Jean's favorite number is an odd number less than 100. The sum of its digits is 9. It has 5 factors.

   a. What is Mina's favorite number?

   b. What is Jean's favorite number? Explain how you found Jean's favorite number.

MILE 8: EXPONENTS

Exponents are a way of writing repeated multiplication. In this mile, you'll practice using exponents in different kinds of expressions. Numbers that are multiplied together are called factors. When two or more factors are the same, you can simplify the expression by using exponents.

Example 1: $2 \times 2 \times 2 \times 2 \times 2 \times 2 = 2^6 = 64$

In the example above, 6 is the exponent and 2 is the base. The expression 2^6 tells you to multiply 2 by itself 6 times. Do you see how writing 2^6 is much simpler than writing out all of those 2s?

Another word that comes up when dealing with exponents is *power*. An exponent indicates the power of an expression. If you refer back to the first example, 2^6, you can say this expression is "two to the sixth power."

Sometimes variables—letters or symbols that represent a certain number—may have exponents too. Treat them just like regular numbers. For example, $y \times y$ is equal to y^2, and $y \times y \times y \times y$ is equal to y^4.

Example 2: $10y^3 = 5 \times 2 \times y \times y \times y$

The example above shows you all the factors of the expression $10y^3$. The number 10 is multiplied by y^3. The factors of 10 are 5 and 2, so the expression is equal to $5 \times 2 \times y^3$. The exponent shows you that there are three ys.

There may be times when you will see a negative number raised to an exponent, such as -4^2. Does this expression mean $-(4 \times 4)$, or $(-4) \times (-4)$? It is important to know because each expression represents a different number. The exponent only applies to the base directly in front of it, so $-4^2 = -(4 \times 4) = -16$. However, if the base has the negative sign contained within the parentheses, such as $(-4)^2$, then that equals $(-4) \times (-4) = 16$.

Example 3: $-7^3 = -(7 \times 7 \times 7) = -343$ $(-7)^3 = (-7) \times (-7) \times (-7) = -343$

Notice that in the example above the answers are the same. When you have an odd exponent for a negative base, you have an odd number of negative factors. The product will therefore be negative. When you have an even exponent for a negative base, you have an even number of negative factors, so the product will be positive. For example, $(-4)^2 = 16$.

You can have a negative exponent. Negative exponents don't make a number negative. They just make the number smaller. To find the value of a negative exponent, you just figure out the inverse of the positive exponent.

Example 4: $3^2 = 9$ $3^{-2} = \dfrac{1}{9}$ $3^3 = 27$ $3^{-3} = \dfrac{1}{27}$

Remember that an integer has 1 as its denominator (so 9 is the same thing as $\frac{9}{1}$). Simply flip the fraction of a positive exponent, and you'll end up with the result of the negative exponent. It's one way to find the value of a negative exponent.

Directions: Match each exponential expression in the first column to its equivalent expression in the second column. If you match all of the items correctly, you will spell out the answer to the question at the bottom of the page.

1. _____ 9^6

2. _____ y^3

3. _____ $(2 + 3)^2$

4. _____ $y^2 \times y^2$

5. _____ $5y^4$

6. _____ $4y^2$

7. _____ 6^9

8. _____ $12y^2$

9. _____ 5^{-3}

10. _____ y^7

11. _____ $9^6 \times 9^2$

12. _____ 6^{-2}

13. _____ $2 \times y \times y$

14. _____ 4^3

15. _____ $6y^3$

16. _____ 4^4

R) 5^2

T) 6 to the ninth power

E) $3 \times 4 \times y \times y$

U) $6 \times y \times y \times y$

R) $\frac{1}{125}$

T) $2y^2$

N) $\frac{1}{36}$

E) $5 \times y \times y \times y \times y$

E) 9^8

R) $4 \times 4 \times 4$

O) y to the third power

C) $y \times y \times y \times y$

W) 9 to the sixth power

C) $y^5 \times y^2$

S) $2 \times 2 \times y \times y$

M) 256

What is a fun place to see a concert in Massachusetts?

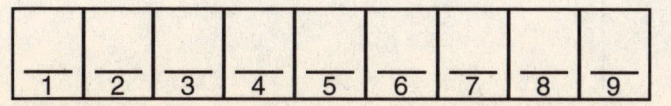
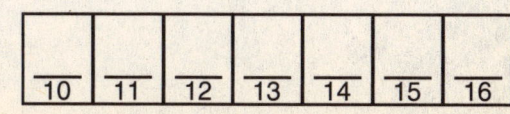

Directions: Read each question carefully. Write your answer in the space below the question or circle the letter next to your answer choice.

1 Which of the following is another way to write $3y^4$?

A. $3 \times y \times y$

B. $3 \times y \times y \times y$

C. $3 \times y \times y \times y \times y$

D. $y \times y \times y \times y$

2 Which of the following expressions is equivalent to $(7 \times 7 \times 7 \times 7 \times 7 \times 7)^3$?

A. 7^6

B. 7^9

C. 7^{18}

D. 49^7

3 Which of the following is another way to express $n^5 \times n^4$?

A. n^9

B. n^1

C. n^{20}

D. n^0

4 Which of the following mathematical expressions shown below is the same as $(5 \times 5 \times 5)^3$?

A. 5^3

B. 5^9

C. 5^6

D. 55^2

5 Which of the following is another way of writing $0.4 \times 300 \times 0.4$ correctly?

A. $300 + 0.4 + 0.4$

B. 304^2

C. 300×0.4^2

D. $300^2 \times 0.4^2$

Tip

When you multiply exponential expressions that have the same bases, you should add the value of the exponents to get the correct product. Be careful not to multiply the values of the exponents! So $4^3 \times 4^2$ is equal to 4^5, not 4^6. Write out the factors to see for yourself that this is true.

6 If $6^2 = 36$, $6^1 = 6$, $6^{-1} = \frac{1}{6}$, and $6^{-2} = \frac{1}{36}$, what does 6^{-3} equal?

   A. $\frac{1}{216}$

   B. $\frac{1}{72}$

   C. $\frac{1}{42}$

   D. $\frac{1}{18}$

7 What is the value of $(11 - 7)^3$?

8 What is the value of $(-3)^3$?

   A. -27

   B. -9

   C. 9

   D. 27

9 Evaluate the expression below.

   $(-6)^2$

10 What is the value of the expression below?

    -2^4

    A. -16

    B. -8

    C. 8

    D. 16

11 What is the value of the expression below?

    2^{-3}

12 What is the value of the expression below?

    5^{-2}

13 How can you write $3^4 \times 3^3$ using a single exponent?

MILE 9: ORDER OF OPERATIONS

When evaluating expressions with more than one operation, you must follow a certain order. The order of operations can be thought of as four steps. First, perform any operations within parentheses. Then, simplify the exponents. Next, multiply or divide in order from left to right. Finally, add or subtract, in order from left to right.

Remember this phrase when doing a problem with multiple operations: "Please Excuse My Dear Aunt Sally," also known as PEMDAS. This will help you follow the order of operations when calculating a problem. P = **p**arentheses, E = **e**xponents, MD = **m**ultiplication and **d**ivision, and AS = **a**ddition and **s**ubtraction.

Example 1: $\left[(6 \times 8) + 4\right] \div 8 + 1.5 = 8$

When you start this problem, you will notice that there is a set of brackets and a set of parentheses. Brackets act like parentheses. Start with $6 \times 8 = 48$ and then add 4 to it to get 52. After that, divide by 8 to get 6.5, and only then add the 1.5 to get the answer, 8.

Example 2: $5^3 \div 15 = 8.\overline{3}$

According to PEMDAS, you should solve for the exponents before you divide. Remember that 5^3 has the same value as $5 \times 5 \times 5$, which is equal to 125. When you divide 125 by 15 you get a number that has a repeating decimal. Don't forget that a repeating decimal should have a bar over the last number.

Example 3: $7 \times 6 - (11 - 8)^3 = 15$

Follow PEMDAS and solve the math in the parentheses first: $11 - 8 = 3$. The exponent gets solved next, so $3^3 = 3 \times 3 \times 3$, or 27. Now you've got $7 \times 6 - 27$. Multiplication gets solved before subtraction, so $7 \times 6 = 42$ and $42 - 27 = 15$.

Directions: Read the riddle below. To find the answer to the riddle, solve each math problem. The answers to the math problems are listed in the boxes at the bottom of the page. In each box, write the letter that corresponds to the answer in the box. If an answer appears in more than one box, you'll have to write the letter more than once.

It goes through an apple,
it points out the way,
it fits in a bow,
then a target, to stay.

What is it?

1. $10 \times 6 + -20 =$ _____ $= A$

2. $-6(-6 + 4) =$ _____ $= O$

3. $\frac{8}{9} \times \frac{6}{7} =$ _____ $= W$

4. $[(6 \times 11) - 9] \div 5 + 8 =$ _____ $= N$

5. $3^2 + 4 =$ _____ $= R$

Answer:

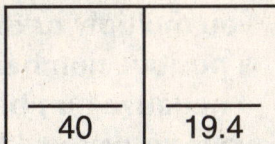
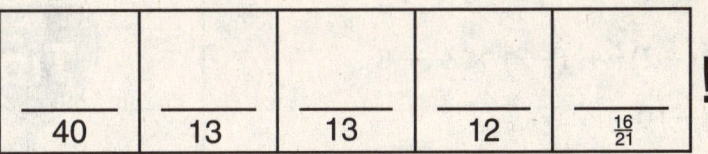

40	19.4

40	13	13	12	$\frac{16}{21}$

!

Directions: Read each question carefully. Write your answer in the space below the question or circle the letter next to your answer choice.

1   What is the value of the expression below?

$$3 + (-2)^3 - (-4) \div 2$$

A. −7

B. −3

C. 9

D. 13

2   What is the value of the expression below?

$$3 + (-2)^2 - (-4) \times 2$$

A. −16

B. −1

C. 1

D. 15

3   What is the value of the expression below?

$$(-2)^2 + (-2) \times 3 - 8$$

A. −10

B. −2

C. 2

D. 10

4   Evaluate the expression below.

$$8 - 2 + 3^2 \div (-3)$$

A. −5

B. −3

C. 3

D. 5

5   Evaluate the following expression:

$$(2)^2 + 4 \times (-2) - 2$$

A. −18

B. −12

C. −6

D. −2

6   What is the value of the expression below?

$$(-6 + 3 \div 3) - 2$$

A. −7

B. −3

C. 7

D. 9

Remember that when you multiply or divide a positive number by a negative number, your answer will be a negative number. When you multiply or divide two negative numbers, your answer will be a positive number.

7 What is the value of the expression below?

$$[(3 + (2)^3 - (-5)] \times 2$$

A. −32

B. −6

C. 21

D. 32

8 What is the value of the expression below?

$$2 \times (-5 + 2 \times 2)$$

A. −12

B. −6

C. −2

D. 24

9 Evaluate the following expression:

$$-7 + 3 \times (2 - 6)$$

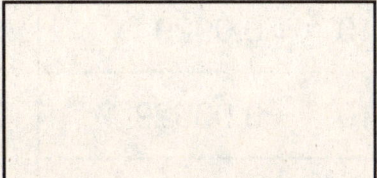

10 Evaluate the following expression:

$$5 \times (-5) + (1 + 4 \times 2)$$

11 What is the value of the expression below?

$$3^2 - 2 \times (-8) + 3$$

12 What is the value of the expression below?

$$4 + (-2) \times 5 - 8 + (-3)^3$$

13 Evaluate the following expression:

$$5 \div (-5) + [(-1) + 4 \times 2]$$

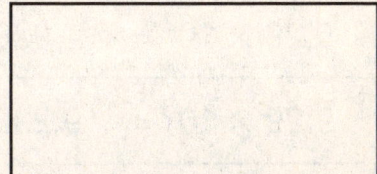

14 Evaluate the following expression:

$$-3^2 - 4 \div 2 + 13$$

MILE 10: SCIENTIFIC NOTATION

Scientific notation is a method of writing very large and very small numbers. This notation shortens a very large or very small number to just a few digits and decreases the chance of leaving out a zero or misplacing the decimal point.

A number in scientific notation is written as a product of a factor and a power of 10. The factor *must* be greater than or equal to 1 and less than 10.

Example 1: $5{,}340{,}000{,}000 = 5.34 \times 10^9$

To write a large number in scientific notation, move the decimal point to the left until you get a number greater than or equal to 1 and less than 10. The number of places you move is equal to the exponent of 10.

$$5.340000000.$$

Example 2: $0.00003624 = 3.624 \times 10^{-5}$

To write a small number in scientific notation, move the decimal point to the right until you get a number greater than or equal to 1 and less than 10. The number of places you move is equal to the negative exponent of 10.

$$.00003.624$$

When you have a number in scientific notation and you want to rewrite it as a decimal, look at the exponent of 10. If it is negative, move the decimal that many places to the left. If it is positive, move the decimal that many places to the right. Use zeros to hold place values.

Study the chart below for more examples.

$9.29 \times 10^1 =$ 92.9		$9.29 \times 10^{-1} = 0.929$
$9.29 \times 10^2 =$ 929		$9.29 \times 10^{-2} = 0.0929$
$9.29 \times 10^3 =$ 9,290		$9.29 \times 10^{-3} = 0.00929$
$9.29 \times 10^4 =$ 92,900		$9.29 \times 10^{-4} = 0.000929$
$9.29 \times 10^5 = 929{,}000$		$9.29 \times 10^{-5} = 0.0000929$

Scientific notation gets its name because *it's actually used in science.* For example, the distance from Earth to the Sun is approximately 93,000,000 miles. This distance can be expressed in scientific notation as 9.3×10^7 miles.

Directions: Convert the numbers below from scientific notation to numbers in their standard form. Then place the answers in the corresponding places in the cross-number puzzle below. Each decimal point belongs in its own box in the puzzle. Also, each answer that includes a decimal point should have a zero before it. For example, ".4" should be written as "0.4" in the puzzle. Do not write in any commas or decimals that come at the end of numbers.

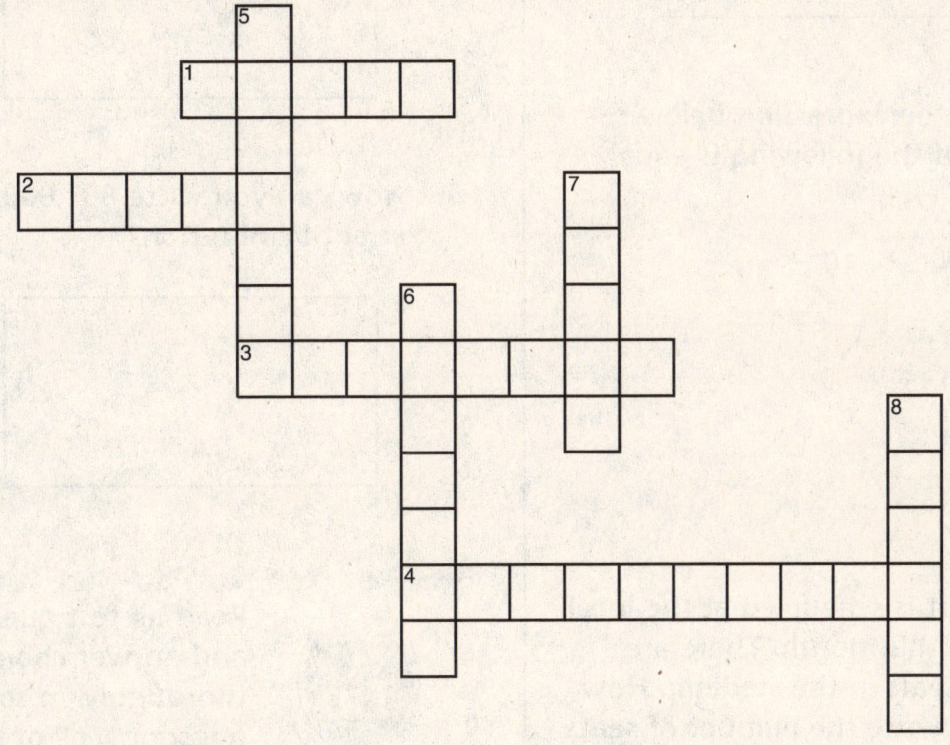

Across

1. 8.96×10^{-1}

2. 1.868×10^4

3. 7.019×10^7

4. 4.823×10^{-5}

Down

5. 3.67×10^{-3}

6. 2.9×10^6

7. 7.75×10^4

8. 3.51×10^{-2}

Directions: Read each question carefully. Write your answer in the space below the question or circle the letter next to your answer choice.

1 How can you express 0.00843643 in scientific notation?

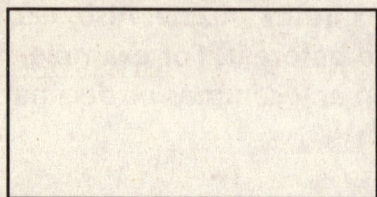

2 Look at the expression below. Which of the following is equal to it?

$$345.435 \times 10^{-2}$$

A. 0.345

B. 3.45435

C. 3,454

D. 34,543.5

3 A concert is scheduled at the local stadium this month. There are 24,000 seats in the stadium. How can you write the number of seats in scientific notation?

A. 240×10^3

B. 2.40×10^3

C. 2.4×10^{-4}

D. 2.40×10^4

4 A calculator showed the answer to a problem as 134,000. How can this number be written in scientific notation?

5 How can you write 801,000,000 in scientific notation?

Read all test questions and answer choices thoroughly. In some questions, all of the details will be important in choosing the correct answer. Always read all four answer choices before you pick one.

6   Tickets to all seven games of the hockey league's finals sold out in record time. 304,406 tickets were sold. How can you express this number in scientific notation?

A. 3.04406×10^{-2}

B. 3.04406×10^5

C. $30,406 \times 10^2$

D. $306,406 \times 10^3$

7   How can you write the quotient of the expression below in scientific notation?

$25\overline{)5,500}$

A. 2.2×10^2

B. 2.2×10^{-2}

C. 22×10^4

D. 22×10^{-1}

8   Which of the following is 4.543×10^3 equal to?

A. 0.004543

B. 0.04543

C. 454.3

D. 4,543

9   How can you express 0.0053446 in scientific notation?

10   New York City and Denver, Colorado, are approximately 1,800 miles apart. If you were to express this distance in scientific notation, how would it look?

A. 1.8×10^{-3}

B. 1.8×10^3

C. 1.8×10^{-4}

D. 1.8×10^4

11   How can you write the quotient of $2,160 \div 4,800$ in scientific notation?

A. 4.5×10^{-1}

B. 45×10^{-1}

C. 4.5×10^{-2}

D. 4.5×10^{-4}

MILE 11: SQUARE ROOTS

The **square root** of a number is one of its two equal factors. That means a square root multiplied by itself gives you the original number.

Example 1:   **What is the square root of 9?**

The square root of 9 is 3 because 3×3 equals 9. The two factors are equal and yield the given number when multiplied together. You can represent this as $\sqrt{9} = 3$. Square roots are related to square numbers, so you can also think of this as $3^2 = 9$.

A **perfect square** is a number that has a whole number as its square root. The example above, 9, is a perfect square because its square root, 3, is a whole number.

You can represent perfect squares as squares of dots. Observe the pattern of dots below. The total number of dots in each square is a perfect square. Follow the pattern of adding dots to make the next perfect square in the space on the right.

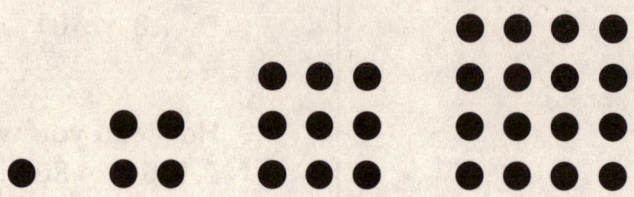

Not all square roots come out so evenly, however. In fact, most numbers are not perfect squares and do not have whole number square roots. In these cases you can approximate the square root of a number by finding the perfect squares that are close to the number.

Example 2:   **Between which two whole numbers does $\sqrt{113}$ lie?**

The number 113 is not a perfect square. But do you know of any perfect squares that are close to 113? The perfect squares 100 and 121 are pretty close. The square root of 100 is 10 and the square root of 121 is 11. Because the number 113 is between the perfect squares 100 and 121, the square root of 113 is somewhere in between 10 and 11. If you find the square root of 113 on a calculator, you'll see that it is approximately 10.63.

Notice that 113 is a little more than halfway between 100 and 121 and its square root is a little more than halfway between 10 and 11. What do you think the square root of 105 is? 105 is between the same two perfect squares, 100 and 121, so you know its square root is between the square roots of those numbers. But because the number 105 is closer to 100, its square root will be closer to 10. In fact, if you find $\sqrt{105}$ on a calculator, it is approximately 10.25.

Directions: Use pencil and paper or a calculator to solve the square root problems at the bottom of the page. Then place the answers in the corresponding places in the cross-number puzzle below. Each decimal point belongs in its own box in the puzzle. Also, each answer that includes a decimal point should have a zero before it. For example, ".4" should be written as "0.4" in this puzzle.

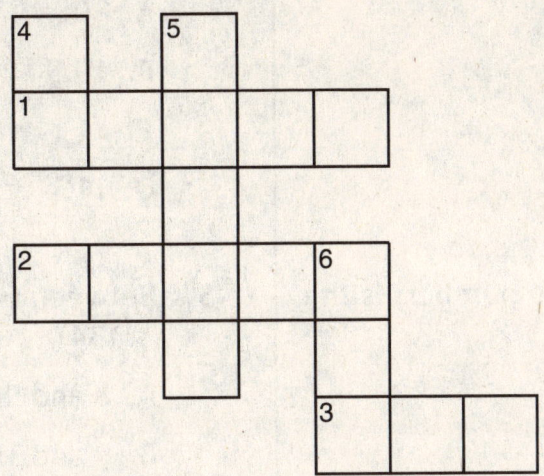

Across

1. What is the value of $\sqrt{311}$ to the hundredths place?

2. What is $\sqrt{34}$ rounded to the thousandths place?

3. What is $\sqrt{0.81}$?

Down

4. What is $\sqrt{121}$?

5. What is the value of $\sqrt{90}$ rounded to the thousandths place?

6. What is one of the two equal factors of 10,000?

Directions: Read each question carefully. Circle your answer choice for each question.

1   Between which two numbers is the square root of 44?

   A. 4 and 5

   B. 5 and 6

   C. 6 and 7

   D. 7 and 8

2   Between which two numbers is the square root of 72?

   A. 8 and 9

   B. 9 and 10

   C. 10 and 11

   D. 11 and 12

3   Which of the following is not a perfect square?

   A. 36

   B. 49

   C. 64

   D. 80

4   Which of the following best represents the square root of 111?

   A. 10

   B. 10.5

   C. 11

   D. 11.1

5   Between which two numbers is $\sqrt{134}$?

   A. 8 and 9

   B. 9 and 10

   C. 10 and 11

   D. 11 and 12

Tip

Square root questions may be asked with or without the square root symbol, $\sqrt{}$, also called a radical. Sometimes a question simply asks for the square root of a number.

6 Which letter on the number line best represents the square root of 56?

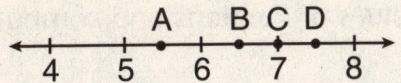

A. A

B. B

C. C

D. D

7 Which of the following is the best estimate for $\sqrt{66}$?

A. 8.1

B. 8.8

C. 9.2

D. 9.4

8 What is $\sqrt{144}$?

A. 14.4

B. 11

C. 14

D. 12

9 Which of the following is the best estimate for $\sqrt{30}$?

A. 4.1

B. 5.5

C. 6.2

D. 6.4

10 Which of the following is the best estimate for $\sqrt{79}$?

A. 8.1

B. 8.9

C. 9.4

D. 9.7

11 What is $\sqrt{169}$?

A. 13

B. 14

C. 15

D. 16

MILE 12: FRACTIONS

In this mile, you will get some practice comparing the values of fractions and computing with fractions.

A fraction is a rational number that is not an integer and is in the form $\frac{a}{b}$, where bmn is not zero. In fractions, a is the numerator and b is the denominator. Fractions show parts of a whole.

Example 1: **The following are the measurements for ingredients for a recipe, in cups. Order them from least to greatest.**

$$1\frac{2}{5}, \ 1\frac{2}{3}, \ 1\frac{4}{5}, \ 1\frac{1}{6}$$

To order fractions you need to have a common denominator. You can find a common denominator for the fractions and then make them into equivalent fractions. The common denominator for these fractions is equal to the lowest common multiple for the denominators, 30. The equivalent fractions are as follows: $1\frac{2}{5} = 1\frac{12}{30}$, $1\frac{2}{3} = 1\frac{20}{30}$, $1\frac{4}{5} = 1\frac{24}{30}$, and $1\frac{1}{6} = 1\frac{5}{30}$. Now you can order them from least to greatest: $1\frac{1}{6} < 1\frac{2}{5} < 1\frac{2}{3} < 1\frac{4}{5}$.

A trick to ordering fractions without having to find equivalent fractions is to compare only the denominators or the numerators. If the denominators are the same, then the fraction with the greater numerator is greater. Compare $1\frac{2}{5}$ and $1\frac{4}{5}$: $1\frac{4}{5}$ is greater because 4 is greater than 2. If the numerators are the same, however, then the fraction with the smaller denominator is greater. So for $1\frac{2}{5}$ and $1\frac{2}{3}$, $1\frac{2}{3}$ is greater because the numerators are equal and 3 is the smaller denominator. This makes sense because the whole is broken up into fewer parts, so each of those parts is bigger. Sometimes, you may be able to order all the fractions using this technique, or eliminate answers this way.

Example 2: **What is the value of $2\frac{2}{3} + 2\frac{1}{2}$?**

In order to add or subtract fractions, you need to rewrite the fractions so they have the same denominator. The common denominator for the fractions in this problem is 6, so $2\frac{2}{3} + 2\frac{1}{2}$ becomes $2\frac{4}{6} + 2\frac{3}{6}$. Add the whole numbers, and then add the numerators to get $4\frac{7}{6}$, or $5\frac{1}{6}$.

Example 3: What is the value of $3 \div \frac{1}{4}$?

To divide fractions, you multiply by the reciprocal of the divisor. The reciprocal of $\frac{1}{4}$ is $\frac{4}{1}$, so you can rewrite $3 \div \frac{1}{4}$ as $3 \times \frac{4}{1}$. To multiply fractions, just multiply the numerators and then multiply the denominators: $\frac{3}{1} \times \frac{4}{1} = \frac{12}{1}$, or simply 12.

Directions: Solve each of the problems below using your knowledge of fractions. Each answer corresponds to a letter on the number line below. As you solve each problem, place the letter in the corresponding box at the bottom of the page. If you match the numbers and letters correctly, you will figure out the answer to the riddle at the bottom of the page.

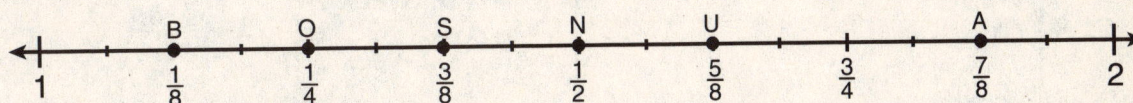

1 Which fraction is greater than $1\frac{3}{4}$, but less than $1\frac{15}{16}$? _____

2. Which fraction is less than $1\frac{1}{6}$? _____

3. Which fraction has the same value as $\frac{1}{2} \div \frac{2}{5}$? _____

4. Which fraction has the same value as $\frac{3}{4} \div \frac{1}{2}$? _____

5. Which fraction has the same value as $2\frac{1}{2} - \frac{7}{8}$? _____

6. Which fraction has the same value as $1\frac{9}{24}$? _____

What do you get for extra credit?

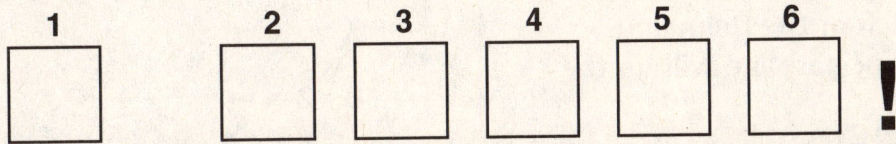

Directions: Read each question carefully. Circle your answer choice for each question.

1. Cynthia has box with a length of $8\frac{3}{5}$ cm. She also has four objects of different lengths. Which one of the following length objects will fit into this box?

   A. $8\frac{7}{8}$ cm

   B. $8\frac{1}{2}$ cm

   C. $8\frac{4}{5}$ cm

   D. $8\frac{9}{10}$ cm

2. Kurt, Gail, Trish, and Jose each had the same amount of modeling clay to use in their art projects. Kurt used $\frac{2}{5}$ of his clay, Gail used $\frac{3}{7}$ of her clay, Trish used $\frac{7}{12}$ of her clay, and Jose used $\frac{2}{3}$ of his clay. Who used the most clay?

   A. Gail

   B. Jose

   C. Kurt

   D. Trish

3. Four identical cars have the same gas mileage and tank size. The car with which of the following amounts of gasoline will go the farthest?

   A. $\frac{3}{5}$ tank

   B. $\frac{2}{3}$ tank

   C. $\frac{7}{9}$ tank

   D. $\frac{5}{12}$ tank

4. Yasna has a spelling test each week. She had $\frac{3}{5}$ correct the first week, $\frac{4}{9}$ correct the second week, $\frac{7}{10}$ correct the third week, and $\frac{5}{9}$ correct the fourth week. Which week did she score the lowest?

   A. week 1

   B. week 2

   C. week 3

   D. week 4

5. Ashley spends $\frac{1}{3}$ of her day sleeping, $\frac{1}{4}$ of her day doing schoolwork, $\frac{1}{12}$ of her time practicing the piano, and $\frac{2}{15}$ of her time at the gym. Which activity does Ashley spend the most time doing?

   A. gym

   B. piano

   C. schoolwork

   D. sleep

6. Which of the following expressions has the same value as $2 \div \frac{1}{3}$?

   A. 2×3

   B. $2 \times \frac{1}{3}$

   C. $\frac{3}{2}$

   D. $\frac{2}{3}$

7 Which of the following expressions has the same value as $a \div \frac{1}{100}$?

A. $0.01a$

B. $10a$

C. $\frac{100}{a}$

D. $100a$

8 Which of the following expressions has the same value as $3 \div \frac{p}{2}$?

A. $\frac{6}{p}$

B. $\frac{p}{6}$

C. $6p$

D. $60p$

You can quickly eliminate some answer choices when ordering fractions. Compare fractions with the same numerators, and then compare fractions with the same denominators. You just might save yourself from doing some messy calculations!

9 Which of the following expressions has the same value as $5x$?

A. $\frac{5}{x}$

B. $5 \div \frac{1}{x}$

C. $x \div 5$

D. $x \bullet \frac{1}{5}$

10 Which of the following expressions has the same value as $t \div \frac{10}{1}$?

A. $10t$

B. $t \times \frac{1}{10}$

C. $t \div \frac{1}{10}$

D. $\frac{10}{t}$

11 Which of the following expressions has the same value as $\frac{b}{100}$?

A. $100b$

B. $\frac{100}{b}$

C. $b \times \frac{1}{100}$

D. $b \div \frac{1}{100}$

MILE 13: DECIMALS

A **decimal** is a number that includes a decimal point. Everything to the right of the decimal point shows part of a whole. So the number 3.78 shows 3 wholes and 78 hundredths.

You can perform operations on numbers with decimals just like you do on whole numbers. When adding or subtracting decimal numbers, line up the decimal points. When multiplying decimal numbers, count the number of decimal places in the factors and make sure that you have the same number of decimal places in the product. When dividing decimal numbers, move the decimal point in the divisor to the right to clear the divisor of any decimals, and then move the decimal point in the dividend the same number of places to the right.

Whew! Sounds complicated, but it just takes some practice. Look at some examples.

Example 1: **Julia had 1.3 m of fabric. She will use 0.25 m for a project. How much fabric will Julia have left?**

This question requires you to subtract. First, rewrite the numbers so that both numbers have the same number of decimal places. To do this, place a zero placeholder in the hundredth place for 1.3. This way the decimal points will line up. The problem becomes

$$\begin{array}{r} 1.30 \\ -\ 0.25 \\ \hline \end{array}$$

Subtract, making sure the decimal point is in your answer: $1.30 - 0.25 = 1.05$ m.

Example 2: **What is $4.32 \div 1.2$?**

First, rewrite the problem with the long division sign so that you can work on it. There is a decimal in the divisor 1.2, so move the decimal one place to the right to clear it. Then, move the decimal point in the dividend, 4.32, one place to the right also. The expression $4.32 \div 1.2$ is the same as $43.2 \div 12$. Now divide, making sure to line up the decimal point in the quotient, to get 3.6.

$$\begin{array}{r} 3.6 \\ 12\overline{)43.2} \\ -36\downarrow \\ \hline 72 \\ -72 \\ \hline 0 \end{array}$$

Example 3: **0.3×0.4**

Multiply the two factors as you would a whole number: $3 \times 4 = 12$. Then count the number of decimal places in the factors and make sure your product has the same number of decimal places. 0.3 has one decimal place and 0.4 also has one decimal place, so the product should also have two decimal places: $0.3 \times 0.4 = 0.12$.

Directions: Read each question carefully. Circle your answer choice for each question.

1 A bench in a garden is placed on a square concrete slab. The concrete slab is 0.8 m in length. What is the area of the concrete slab, in square meters?

A. 0.064

B. 0.64

C. 6.4

D. 64

2 What is the value of (0.3)(0.3)(0.3)(0.3)?

A. 0.0081

B. 0.081

C. 0.81

D. 8.1

3 What is the value of (0.2)(0.2)(0.2)?

A. 0.0008

B. 0.008

C. 0.08

D. 0.8

4 What is the value of (0.4)(0.4)?

A. 0.0016

B. 0.016

C. 0.16

D. 1.6

5 Carole paid $15.75 for 3.5 pounds of peanuts. How much did Carole pay for each pound of peanuts?

A. $0.05

B. $0.45

C. $4.50

D. $45.00

When you multiply decimals, the total number of decimal places in the factors should equal the number of decimal places in the product. You can immediately eliminate answer choices with the incorrect number of decimal places in the product.

MILE 14: PERCENTS

Percent means "per hundred" or "out of 100." It shows how much something is out of 100. Like decimals or fractions, percents show parts of wholes. Understanding percents can help you to figure out discounts or selling prices.

Example 1: **24 out of 100 cars on a highway are blue. What percent of the cars are blue?**

A percent is the part out of 100. Because 24 out of 100 cars are blue, you can determine that 24% of the cars are blue.

Example 2: **2 out of every 5 students take Spanish. What percent of the student body takes Spanish?**

Although there may be fewer than 100 students, you can still find the percent. First, write the fraction for the number of students that take Spanish, which is 2 out of 5, or $\frac{2}{5}$. Then, to find the percent, write an equivalent fraction with a denominator of 100. Multiply the numerator and denominator by 20 so that the denominator equals 100. $\frac{(2 \times 20)}{(5 \times 20)} = \frac{40}{100}$. The percent of students who take Spanish is 40%.

Example 3: **An item that regularly sells for $12 is discounted 10%. What is the new price of the item?**

A discount is an amount that is taken off the original price. To find the discounted amount, you can rewrite the percent as a fraction: 10% is 10 out of 100, or $\frac{10}{100}$. So the discount is $\frac{10}{100}$ of the original price, or $\frac{10}{100} \times \$12 = \$1.20$. Remember that $1.20 is to be taken off the original price of $12. So the new price is $12.00 - $1.20, or $10.80.

Example 4: **30% of what number is 24?**

Sometimes you can write an equation to make things easier when doing percent problems. Use a variable for the unknown number. You can use the variable x. You know that 30% is $\frac{30}{100}$ and "of" indicates multiplication. Now you can write the equation $\frac{30}{100} \cdot x = 24$, and solve for x. To isolate and find x, multiply both sides by 100 and divide by 30. You will get $x = 24 \times 100 \div 30 = 80$. So 30% of 80 is 24.

Directions: Match the problem in the left column with the best answer in the right column. Each answer should match only one problem.

1. $\frac{3}{25}$ as a percent   9

2. 20% of 45   12%

3. 30% × 40   6

4. 24 out of 60 as a percent $33\frac{1}{3}$%

5. 14% of 50   3.3

6. $\frac{1}{3}$ as a percent   40%

7. 15% of 40   225%

8. 22% of 15   7

9. 25% of 14   3.5

10. 45 out of 20 as a percent   12

Directions: Read each question carefully. Write your answer in the space below the question or circle the letter next to your answer choice.

1. A bike shop has bikes for 40% off the regular price. What is the sale price of a bike that regularly sells for $90?

   A. $36

   B. $54

   C. $80

   D. $126

2. A pair of shoes that normally sells for $40 is discounted 15%. What is the sale price of the shoes?

   A. $6

   B. $25

   C. $34

   D. $46

3. The cost of a calculator is 20% off the suggested retail price. If the suggested retail price is $15, what is the cost of the calculator?

   A. $12

   B. $13

   C. $14

   D. $18

4. The liquid in a pot is boiled until it is reduced by 10%. If there were 10 cups of liquid before boiling began, how much liquid was left in the pot?

   A. 1 cup

   B. 5 cups

   C. 7 cups

   D. 9 cups

5. There were 70 students trying out for a school's football team. The coach had to reduce this number by 30% to make the final team. How many students will be on the final team?

6. The 200 eighth-grade students at a school were invited to visit the local high school. Ninety percent of the eighth graders attended. How many eighth graders visited the high school?

7 The quality control department at a toy factory found that 20% of their toys were defective. The factory inspected a total of 300 toys. How many of these toys were defective?

8 40% of what number is 10?

9 70% of what number is 21?

10 What number is 20% of 35?

A. 7

B. 18

C. 20

D. 28

11 Timothy bought a backpack that was discounted 20%. The regular price of the backpack was $19. About how much did Timothy pay for the backpack?

A. $4

B. $8

C. $12

D. $15

12 20 is about what percent of 30?

A. 20%

B. 40%

C. 70%

D. 90%

MILE 15: FRACTIONS, DECIMALS, AND PERCENTS

In this mile you will compare and order **fractions**, **decimals**, and **percents**. When you compare two fractions, decimals, or percents, you will have to figure out which one is larger or smaller.

It is fairly straightforward to compare and order decimals and percents. You just line up the decimal points and compare each place value from left to right. For example, 3.56 is smaller than 3.65 because the 5 in the tenths place is smaller than the 6 in the other number. You can ignore the digits in the hundredths place because you have already determined which number is larger. Remember to only compare digits in the same places.

Sometimes it is difficult to compare fractions because the numerators and denominators are both different. In these cases you can convert the fractions to decimals and follow the same steps as above. Fractions, decimals, and percents all represent parts of wholes, and you can write equivalent amounts in all three forms.

Knowing how to put rational numbers in order and how to convert among fractions, decimals, and percents will help you figure out everyday things, such as which baseball team has a better record, what percent of a day you spent practicing the piano, or how much money you can earn in interest from a savings account.

Converting with Percents

To convert a decimal to a percent, multiply by 100 and add the % symbol.

$$0.45 \times 100 = 45\%$$

To convert a percent to a decimal, divide by 100 and remove the % symbol.

$$74\% \div 100 = 0.74$$

Converting Between Fractions and Decimals

To convert a fraction, such as $\frac{1}{8}$, to a decimal, divide the numerator by the denominator: $8\overline{)1}$. Remember to add zeros after the decimal point to hold place values. When you are finished dividing you will find that $\frac{1}{8}$ = 0.125. Using the steps above, you can rewrite this decimal as a percent, 12.5%.

To convert 0.4 to a fraction, use the 4 as the numerator and 10 as the denominator: (because the decimal ends at the tenths place). So 0.4 becomes $\frac{4}{10}$. If you'd like, you can simplify $\frac{4}{10}$ to $\frac{2}{5}$.

To convert 0.17 to a fraction, use 17 as the numerator and 100 as the denominator (because the decimal ends in the hundredths place). So 0.17 becomes $\frac{17}{100}$.

Directions: You may need a calculator for this activity. Figure out the answers to the questions listed at the bottom of the page. Then place the answers in the corresponding places in the cross-number puzzle below. Each decimal point belongs in its own box in the puzzle. Each answer that does not include a whole number should have a zero before the decimal point. For example, ".4" should be written as "0.4" in the puzzle.

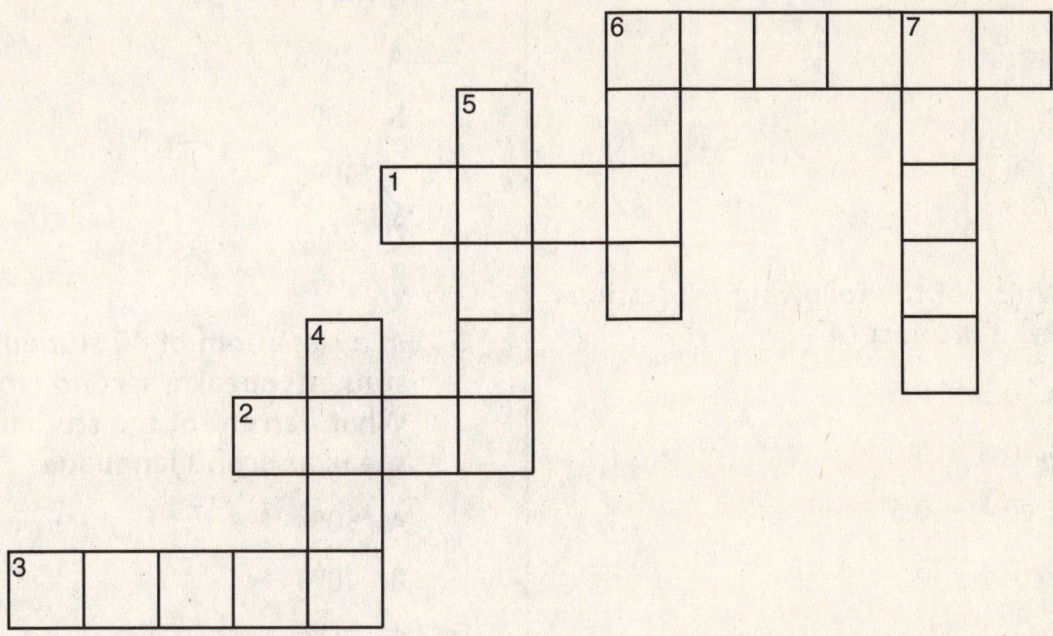

Across

1. The fraction $\frac{27}{4}$ is equal to what decimal?

2. Convert 15% to a decimal.

3. What decimal is expressed as the fraction $\frac{5}{12}$?

6. Express 50.19% as a decimal.

Down

4. Convert $\frac{2}{3}$ to a decimal.

5. The fraction $\frac{3}{8}$ is equal to what decimal?

6. Convert 54% to a decimal.

7. The mixed number $12\frac{3}{4}$ is equal to what decimal?

Directions: Read each question carefully. Circle your answer choice for each question. You may need a calculator to solve some of these problems.

1   Which of the following has the same value as 0.3×0.8?

A. $\frac{6}{50}$

B. $\frac{6}{25}$

C. $\frac{24}{25}$

D. $\frac{8}{3}$

2   Which of the following expressions has a product of $\frac{1}{5}$?

A. 0.3×0.5

B. 0.3×5

C. 0.4×0.5

D. 0.4×5

3   Kia got 19 out of 25 correct answers on her science test. What percent of the answers did Kia get correct?

A. 25%

B. 38%

C. 50%

D. 76%

4   10 out of 40 people at a restaurant ordered dessert.
What percent of the people ordered dessert?

A. 25%

B. 40%

C. 50%

D. 75%

5   In a classroom of 40 students, 30 students speak a second language. What percent of the students speak a second language?

A. 30%

B. 40%

C. 75%

D. 95%

6

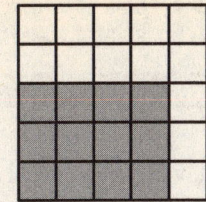

Each square in the diagram above represents a table at a restaurant. If the shaded squares represent the tables that are occupied, what percent of the tables are occupied?

A. 25%

B. 48%

C. 52%

D. 70%

7 14 out of 18 students surveyed own a pet. What percent of those students own a pet, to the nearest percent?

A. 78%

B. 75%

C. 67%

D. 80%

8 In a class of 32 students, 12 ride the bus to school. What percent of the students ride the bus, to the nearest percent?

A. 28%

B. 38%

C. 40%

D. 45%

9 In a science experiment, 12 out of 17 plants survived. About what percent of the plants did not survive?

A. 30%

B. 40%

C. 70%

D. 80%

10 Which of the following has the same value as 0.5 × 0.7?

A. $\frac{1}{35}$

B. $\frac{1}{20}$

C. $\frac{7}{100}$

D. $\frac{7}{20}$

11 Which of the following expressions has a product of $\frac{3}{4}$?

A. 0.3 × 2.5

B. 0.3 × 25

C. 0.3 × 4

D. 0.3 × 40

MILE 16: ESTIMATING

When you **estimate**, you round to quickly figure out an approximate answer to a problem. An *approximate* answer is one that is close to, but is not necessarily, the actual answer.

You should estimate when you do not need an actual answer, such as when you are grocery shopping. If you only have $20 in your pocket, you need to know that the sum of your purchases will be less than the cash in your pocket. You don't need to know the exact sum of your purchases until later at the cash register. You can also estimate when you are checking your work in math. One important reason to use estimation on the MCAS is to eliminate wrong answer choices.

When you estimate, round the numbers into easier numbers to calculate. Then work out the operations. Always ask yourself if your estimate makes sense.

Example 1:   Estimate 28 × 33.

The best way to estimate is to round each of the numbers to the nearest ten. You know that 28 and 33 are both close to 30, so you can multiply 30 × 30 to get a good estimate. Your estimate would be 900. If you took the time to actually multiply 28 × 33 the long way, your answer would be 924. That's pretty close to 900, right?

Here's another trick you can use to estimate multiplication problems: If you round one of the factors to a multiple of 10, such as 10, 20, 30, 40, etc., you can take off the 0 and attach it to the answer after you multiply. So if you multiply 5 × 50, you can think of it as 5 × 5 because those numbers are easier to work with. Multiply 5 by 5 to get 25, and tack on the 0 that you took off in the beginning to get 250. In Example 1, think of 30 × 30 as 3 × 3 with two 0s taken off. So 3 × 3 = 9 with two 0s, or 900!

Example 2:   **Each orange at a fruit stand costs $0.49. If 97 oranges were sold in a day, what is a reasonable estimate of the amount of money the fruit stand made by selling oranges that day?**

To estimate the answer to this problem, you can round $0.49 up to $0.50 and round 97 oranges to 100 oranges. $0.50 × 100 comes out to $50. This is a reasonable estimate because 0.5 is equal to one-half, and half of 100 is 50. Compare this to the actual amount of money that the fruit stand made from selling oranges: $0.49 × 97 = $47.53.

Be careful when you are estimating with decimals or fractions, because they are less than whole numbers. You could have made an error in Example 2 and multiplied 100 by 5 to get $500. This is not reasonable. Just remember to ask yourself if your estimate makes sense.

Directions: Match the problem in the left column with the best estimate in the right column. Each estimate should match only one problem. If you match the items in the two columns correctly, you will spell out the answer to the riddle at the bottom of the page.

1. _____ $\frac{2}{5} + \frac{4}{7}$ W) 4

2. _____ $\frac{1}{3}$ of 57   O) 20

3. _____ 9% × 98   D) 1

4. _____ $\frac{1}{4}$ × 15   D) 6

5. _____ $3\frac{1}{5} - 1\frac{1}{10}$ G) 9

6. _____ 47% of 29   O) 2

7. _____ $2\frac{6}{7} + 2\frac{5}{6}$ O) 15

Answer:

Which dog has a bark, but no bite?

_____

1 About how much is 80% of 96?

A. 20

B. 40

C. 60

D. 80

2 The bookstore is having a one-day sale. All books are 40% off the regular price. What is the best estimate of the sale price of a book that regularly sells for $48?

A. $20

B. $30

C. $40

D. $50

3 Mitsuo has 102 baseball caps in his collection. He has about 10% more caps this year than the number of caps he had last year. About how many baseball caps did Mitsuo have last year?

A. 10

B. 50

C. 90

D. 110

4 An item at a store is on sale for 20% off the regular price. About how much will the item cost, if it regularly sells for $53?

A. $40

B. $50

C. $60

D. $70

5 Rory works part time at a store. He gets a 20% employee discount. If Rory buys an item that sells for $9, about how much will he save on the item?

A. $2

B. $7

C. $9

D. $11

6 Mike's mother has to work 233 days before she gets a paid vacation. If she has already worked $\frac{1}{4}$ of the required days, about how many more days will she have to work before she gets a paid vacation?

A. 60 days

B. 80 days

C. 115 days

D. 180 days

7 Belinda is making bracelets for a craft show. Her goal is to make 123 bracelets. She has completed $\frac{1}{3}$ of the bracelets. About how many bracelets does she still have to make?

A. 40

B. 80

C. 120

D. 160

8 Ingrid had 42 math questions to finish for homework. She has completed about $\frac{1}{4}$ of the questions. About how many questions does she have left?

A. 10

B. 20

C. 30

D. 40

9 Lyle is writing a short story. He figures that the story will be about 14 pages. Lyle has finished $\frac{1}{5}$ of his story. About how many pages has Lyle completed?

A. 1

B. 3

C. 5

D. 12

10 Sue is traveling 289 miles on a train. She has completed about $\frac{1}{2}$ her journey. Estimate how far Sue has traveled.

A. 50 miles

B. 100 miles

C. 150 miles

D. 200 miles

11 Josiah has 17 math problems for homework. It takes him about 5 minutes to do each problem. If Josiah wants to go to his friend's house at 6:30 P.M., approximately what is the latest time he can begin his homework?

A. 5:00 P.M.

B. 5:30 P.M.

C. 6:00 P.M.

D. 6:30 P.M.

12 Brandon reads 28 pages from a book each night before going to bed. It takes him 3 minutes to read each page. If Brandon needs to be in bed by 10:00 P.M., around what time should he begin reading?

A. 8:00 P.M.

B. 8:30 P.M.

C. 9:00 P.M.

D. 9:30 P.M.

RATIOS AND PROPORTIONS

Ratios and proportions are often used in advertising, for scales on a map, and to compare teams in sports.

A **ratio** compares two numbers. You can write a ratio using the symbol : or the word *to*, or as a fraction. The ratio of *x* to *y* can be written $x{:}y$, *x* to *y*, or $\frac{x}{y}$. The order in which you write the ratio matters, so make sure to present the ratio the right way. For instance, $y{:}x$ or $\frac{y}{x}$ is not the same as above. These two ratios mean *y* to *x*.

A **proportion** is a statement where two ratios are equal. The ratios are normally written as fractions. For example, $\frac{2}{4} = \frac{5}{10}$ is a proportion.

Example 1: **In a classroom of 24 students, there are 15 boys. What is the ratio of girls to boys?**

The question asks for the ratio of girls to boys. So the ratio can be written as *girls:boys, girls* to *boys*, or $\frac{girls}{boys}$. Because there are 24 students and 15 are boys, $24 - 15$, or 9 students, are girls. The ratio of girls to boys is $\frac{9}{15}$, which can be reduced to $\frac{3}{5}$. Note that the quantity that is mentioned first is written as the numerator. The ratio of girls to boys is not the same as the ratio of boys to girls, which is $\frac{15}{9}$.

Example 2: **A map is drawn to scale so that 2 centimeters equals 3 miles. If the distance between two cities on a map is 8 centimeters, what is the actual distance between the cities, in miles?**

You can solve this question by setting up a proportion. Write two ratios of the map distance to the actual distance. You know that the scale is 2 centimeters to 3 miles, or $\frac{2 \text{ cm}}{3 \text{ mi.}}$. The second ratio will be 8 centimeters to *d* (the unknown distance), or $\frac{8 \text{ cm}}{d \text{ mi.}}$. Now set the ratios equal to each other: $\frac{2 \text{ cm}}{3 \text{ mi.}} = \frac{8 \text{ cm}}{d \text{ mi.}}$, and solve for the variable, *d*. You can do so by cross multiplying: $2 \times d = 3 \times 8$, or $2d = 24$. So $d = 12$. The actual distance represented by 8 centimeters on the map is 12 miles.

When setting up proportions, make sure that the ratios represent the same thing. In the example above, the ratios were map distance to actual distance. If one of the ratios were switched, then the proportion would be incorrect and you would get the wrong answer. The proportion $\frac{2 \text{ cm}}{3 \text{ mi.}} = \frac{d \text{ mi.}}{8 \text{ cm}}$ is incorrect.

Directions: Read the riddle below. To find the answer to the riddle, solve each math problem. The possible answers to the math problems are listed in the boxes at the bottom of the page. In each box, write the letter that corresponds to the answer in the box. If an answer appears in more than one box, you'll have to write the letter more than once.

1. Express 8 wins to 12 losses as a ratio in simplest form. _____ = A

2. Express 100 miles to 4 gallons as a ratio in simplest form. _____ = T

3. Express 50 cents to 5 gumballs as a ratio in simplest form. _____ = N

4. Express 3 wins in 7 games as a ratio in simplest form. _____ = O

5. Express 10 miles in 3 minutes as a ratio in simplest form. _____ = I

6. What is x in the proportion $\frac{1}{3} = \frac{x}{42}$? _____ = H

7. What is r in the proportion $\frac{4}{7} = \frac{r}{49}$? _____ = E

8. What is b in the proportion $\frac{2}{5} = \frac{6}{b}$? _____ = W

9. What is q in the proportion $\frac{2}{5} = \frac{q}{100}$? _____ = L

10. What is x in the proportion $\frac{x}{3} = \frac{1}{2}$? _____ = S

Why did the duck take a swim around the pond?

Answer:

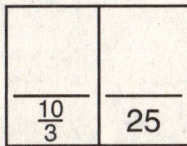
$\frac{10}{3}$ 25

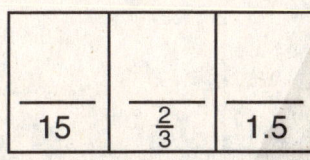
15    $\frac{2}{3}$ 1.5

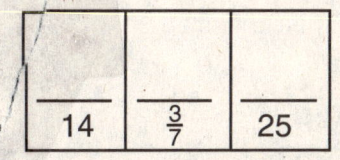
14    $\frac{3}{7}$ 25   !

Directions: Read each question carefully. Write your answer in the space below each question or circle your answer choice for each question.

1   Santiago used 2 cups of chopped nuts and 4 cups of milk to make cookies. What is the ratio of cups of chopped nuts to cups of milk?

   A. $\frac{2}{1}$

   B. $\frac{4}{2}$

   C. $\frac{1}{2}$

   D. $\frac{1}{4}$

2   Edie is 14 years old. Vanessa is 10 years old. What is the ratio of Edie's age to Vanessa's age?

   A. 1 to 14

   B. 1 to 10

   C. 5 to 7

   D. 7 to 5

3   What is the missing term in the proportion below?

   $\frac{4}{2} = \frac{x}{6}$

   A. 6

   B. 12

   C. 18

   D. 24

4   What is the missing term in the proportion below?

   $\frac{8}{10} = \frac{12}{n}$

   A. 15

   B. 20

   C. 25

   D. 30

5

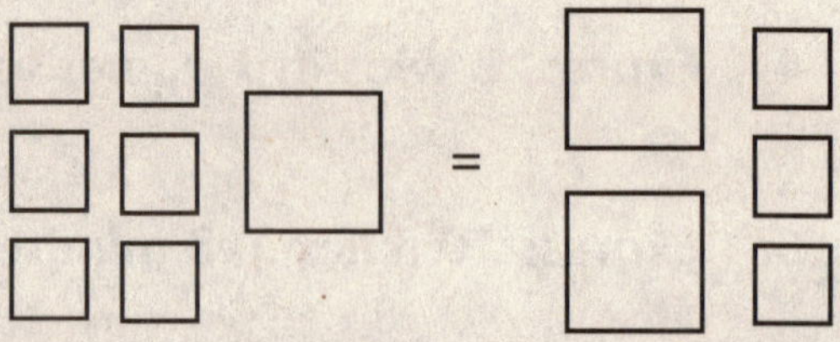

   What is the relationship between a large block and a small block in the diagram above?

   A. One large block is equivalent to two small blocks.

   B. One large block is equivalent to three small blocks.

   C. One small block is equivalent to two large blocks.

   D. One small block is equivalent to three large blocks.

6

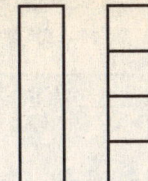

In the diagram above, how many squares make up 4 rectangles?

A. 4

B. 8

C. 12

D. 16

7

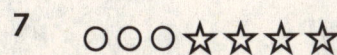

Miguel showed the team's wins and loses using the circles and stars shown above. The circles represent wins and the stars represent losses. What is the ratio of losses to wins?

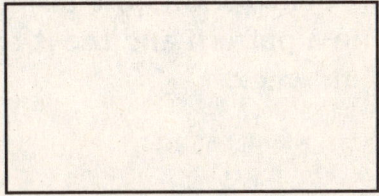

8   What is the ratio of the perimeter, *P*, of a square to one of its sides, *s*?

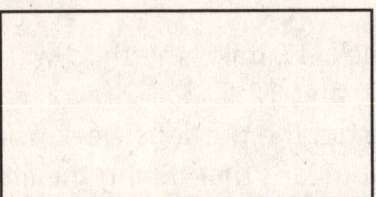

9   The formula for the volume of a prism, *V*, is *Bh*, where *B* is the area of its base and *h* is the height. What is the ratio of the volume, *V*, to the height, *h*, of the prism?

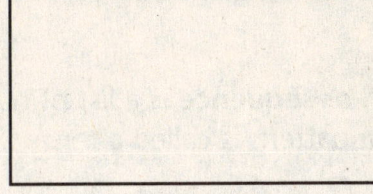

10   What is the ratio of the area of a circle, *A*, to the square of its radius, *r*?

A. $\dfrac{A}{r^2} = \pi$

B. $\dfrac{A}{\pi} = r$

C. $\dfrac{r^2}{A} = \pi$

D. $\dfrac{r^2}{\pi} = A$

Mile 18: Number Patterns

In this mile, you will learn about patterns in numbers and practice extending number patterns. Patterns can help you understand relationships, make connections, and solve problems.

A **pattern** or **sequence** is a list of numbers in a certain predictable order. Each number or figure in a pattern is called a term.

To find a number pattern, look for a relationship between consecutive terms. The relationship has to be predictable and continue for the entire sequence. So if you have to add a specific number to get the next term, that pattern should continue for the rest of the terms.

Example 1:   What are the missing terms in the following pattern?

> 200, 180, 161, 143, 126, _____, _____

You have to figure out the pattern. The terms are getting smaller, so subtract each term from the previous term: 200 – 180 = 20, 180 – 161 = 19, 161 – 143 = 18, 143 – 126 = 17. It appears that the pattern is to subtract the next smaller integer from each term. To get the next term, subtract 16 from 126: 126 – 16 = 110. Then subtract 15 from 110 to get the final term: 110 – 15 = 95. So the missing terms are 110 and 95.

Example 2:   Kenny is training for a bike race. The chart below shows the distances he rides each week during training. Determine a pattern and use it to find the distance Kenny will ride during the sixth week.

Week	Distance
1	3 miles
2	6 miles
3	12 miles
4	24 miles

Kenny rides 3 miles the first week, 6 miles the second week, 12 miles the third week, and 24 miles the fourth week. Because 3 × 2 = 6, 6 × 2 = 12, and 12 × 2 = 24, you can conclude that Kenny rides twice as much each week as he did the previous week. You have to multiply the previous term by 2 to get the next term. That means that during the fifth week he will ride 24 × 2 = 48 miles, and during the sixth week he will ride 48 × 2 = 96 miles.

Directions: Figure out the next number in each pattern below. Then write the numbers into the corresponding spaces in the cross-number puzzle below.

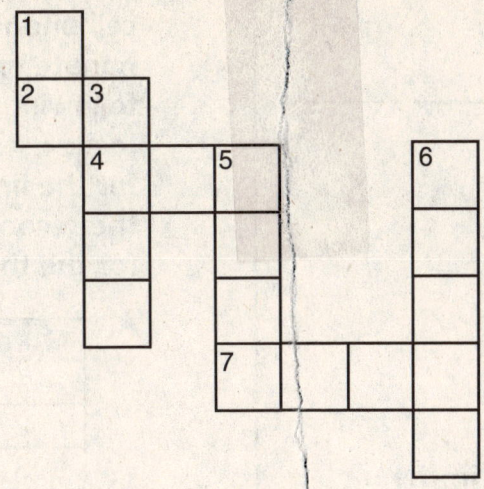

Across

2. What is the sixth term in the pattern 91, 85, 79, 73 . . . ?

4. Name the seventh term in the following pattern: 7, 14, 28, 56, 112 . . .

7. What is the missing term in this pattern? 1,300; 1,600; 2,200; 2,500; 3,100; _____; 4,000

Down

1. What is the next number in the sequence 1, 4, 10, 19, 31 . . . ?

3. What is the next term in the pattern 6, 18, 54, 162, 486 . . . ?

5. Find the missing number in the following sequence: 8,755; 8,753; 8,749; 8,743; 8,735; 8,725; _____.

6. What is the seventh number in the pattern 1,000; 2,000; 4,000; 7,000; 11,000 . . . ?

Directions: Read each question carefully. Circle your answer choice for each question.

1 What are the next three terms in the following pattern?

98, 97, 95, 92, 88, ———, ———, ———

A. 83, 77, 70

B. 84, 80, 76

C. 86, 83, 79

D. 87, 86, 85

2 What is the first term in the following pattern?

———, 4, 8, 16, 32

A. 2

B. 3

C. 4

D. 5

3 What are the next three terms in the following pattern?

1, 2, 4, 7, 11, 16, ———, ———, ———

A. 17, 18, 19

B. 22, 29, 37

C. 26, 36, 46

D. 32, 64, 128

4 Joelle pays $20 a month for her cell phone, and 10 cents for each minute of use. Which of the following tables shows the cost if Joelle uses the phone 10 minutes for the first month, 20 minutes for the second month, and 30 minutes for the third month?

A.

Month	Cost
1	$1
2	$2
3	$3

B.

Month	Cost
1	$21
2	$22
3	$23

C.

Month	Cost
1	$20
2	$30
3	$40

D.

Month	Cost
1	$100
2	$200
3	$300

5 Jennifer borrowed $200 from her brother. She pays him a sum of money each month. The chart below shows the balance that she still owes him after each payment is made.

Month	Remaining Balance
1	$175
2	$150
3	$125

If Jennifer continues to follow the same pattern each month, in which month will the balance be 0?

A. 6

B. 7

C. 8

D. 9

6 Jahleel made the pattern below using marbles. How many marbles will be in the fifth group?

A. 5

B. 16

C. 20

D. 25

7 Ticket prices to the fairgrounds are shown in the table below.

Number of Tickets	Cost
0	$0
2	$1
4	$2
6	$3

If Maria buys 24 tickets, how much will she pay?

A. $6

B. $8

C. $12

D. $24

8 The number of cars sold by a car dealership each year is shown in the table below.

Year	Number of Cars
1999	1,500
2000	2,000
2001	2,500

If the pattern continues, how many cars can the dealership expect to sell in the year 2007?

A. 4,500

B. 5,000

C. 5,500

D. 6,000

MILE 19: GEOMETRIC PATTERNS

Remember that a pattern is a series of numbers or figures in a certain order. The terms in those series are formed by a repeated, predictable pattern. A geometric pattern is a pattern of shapes with a relationship between the shapes.

Look at the shapes below to find a pattern.

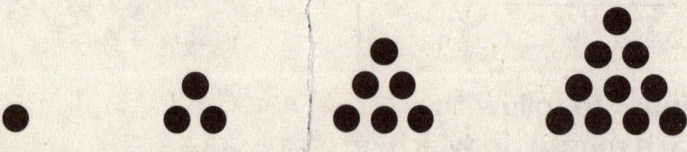

You can see that the pattern is a series of triangles that increase in size. The first shape has only 1 dot. The second shape is made up of 3 dots (2 on each side). The third shape is made up of 6 dots (3 on each side). The fourth shape is made up of 10 dots (4 dots on each side). So the shapes are made up of 1, 3, 6, and 10 dots (with 1, 2, 3, and 4 dots for each side respectively). Each shape has an increasing number of dots on each side. So if the pattern continued, the next shape should have 5 dots on each side, for a total of 5 + 10 or 15 dots.

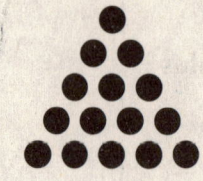

What is the next shape? Draw it below.

Directions: Read each question carefully. Write your answers in the space below each question or on another sheet of paper.

1   The blocks below are arranged in a pattern.

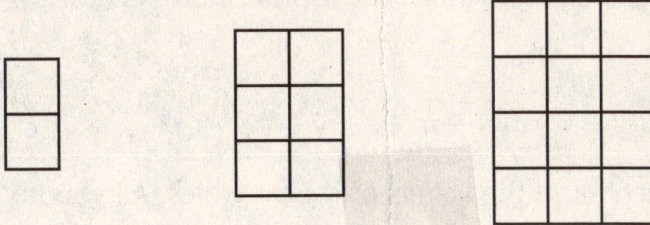

If the pattern continues, how many blocks will be in the fifth pattern?

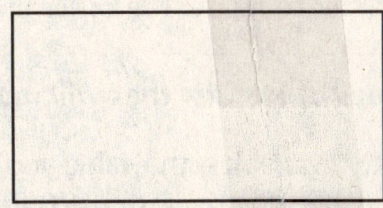

2   A graphic artist made a design for a logo. She created the logos of varying sizes, as shown below.

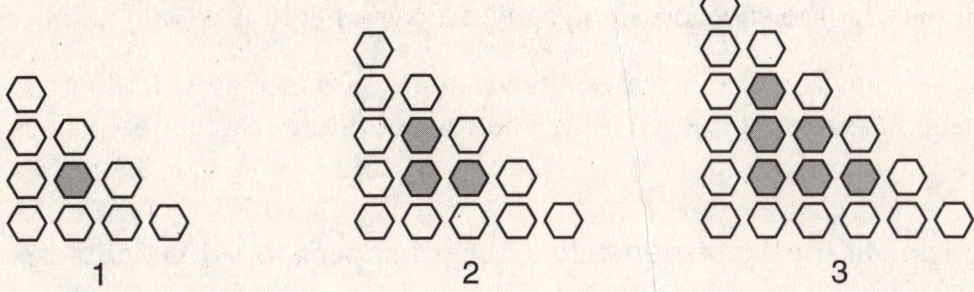

a.   If the pattern continues, how many hexagons will be shaded in the next logo?

b.   Make a table of drawing number to the number of white hexagons.

c.   If the pattern continues, find a rule that will tell you how many white hexagons will be in any future drawings. Explain how you found the rule.

MILE 20: VARIABLES

Variables are symbols or letters that represent unknown numbers. Whenever you are given an algebraic expression to evaluate, you can substitute numbers for the variables to find the result.

Example 1: **If $x = 3$ and $y = 4$, solve for $2x + y$.**

You are given the expression and the values of the variables. All you have to do is substitute the numbers for the variables in the expression and then evaluate. So, replacing x with 3 and y with 4, the expression becomes $2(3) + 4$, or $6 + 4 = 10$.

To evaluate any expression, remember to follow the order of operations, or PEMDAS.

Example 2: **What are some values of x and y that will make $y = 2x + 4$ true?**

When you have an equation with two variables, the value you put in for one of the variables will affect the value of the other variable. If you substitute 0 for x in the equation $y = 2x + 4$, then y is $2 \times 0 + 4$, or 4. If you substitute 2 for x in the equation, y is a different value: $2 \times 2 + 4 = 8$. The value that you substitute for x results in some specific y value. In other words, the value of y depends on the value that you put in for x. Two pairs of x and y values that will solve the equation $y = 2x + 4$ are (0, 4) and (2, 8). There are numerous other pairs that will solve for the equation. Can you find a pair where x is an odd number?

In Example 2, as the value of x increases, the value of y also increases. To find the pattern of how the variables are related to each other you can substitute real number values for the variables.

Example 3: **Look at the expression below. What happens to n if m doubles?**

$$m^2 = n$$

This may look complicated, but you can simplify the problem if you use real numbers. Compare what happens to n when m is equal to 2 and 4. When $m = 2$, $n = 4$. If $m = 4$, then $n = 16$. It looks like n increases by a factor of 4. Use another pair of numbers to check. When $m = 3$, $n = 9$, and when m is doubled to 6, $n = 36$. It looks like when m doubles, n increases by a factor of 4.

It's better to use small numbers as examples. Why make your computations harder than they need to be? Because 1 and 2 have some unique properties when they are used in multiplication and with exponents, use numbers like 3, 4, or 5.

Directions: Match each expression in the first column to its value in the second column.

1. $3a + 6$, when $a = 5$ 25

2. $8 + 2y$, for $y = -2$ 7

3. $(y - 3)^2$, for $y = 6$ -10

4. $y^2 \times y^2$, for $y = 2$ 4

5. $y = x + 4$. When $x = -5$, $y = ?$ 8

6. $4m^2$, when $m = 3$ -1

7. $2 + 3b$, for $b = -4$ 11

8. $\left(\frac{12}{2y}\right)^3$, when $y = 2$ 21

9. $y^5 = x$. When $y = 1$, $x = ?$ 9

10. $5 + \frac{c}{2}$, for $c = 4$ -5

11. $2p + 3 = q$. When $p = 4$, $q = ?$ 16

12. $4x - 5 = y$. When $x = 0$, $y = ?$ 27

13. $a^2 = b$. When $a = 5$, $b = ?$ 36

14. $2^m = n$. When $m = 3$, $n = ?$ 1

Directions: Read each question carefully. Write your answer in the space below each question or circle your answer choice for each question.

1 Evaluate the expression below for $n = 2$ and $m = -3$.

$$5 - [8 - 3(n + 4) - 6m]$$

A. -3

B. 3

C. 5

D. 13

2 What is the value of $-4[8 + 2(a + b) - 2a]$ when $a = 4$ and $b = -2$?

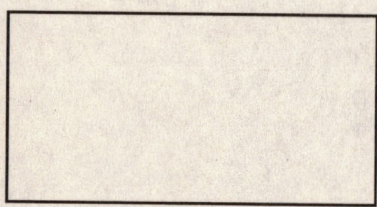

3 What is the value of $3 + 2[8 - 2(mn) - 3n]$ when $m = -1$ and $n = 3$?

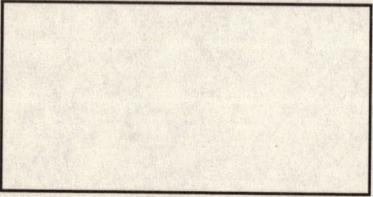

4 Evaluate $3[4 + 6(a - b) + b]$ when $a = 3$ and $b = 3$.

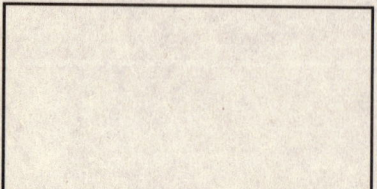

5 Which of the following are possible values for d and g in the equation below, if d is an odd number between 0 and 5?

$$g = 5 - d$$

A. $d = 2, g = 3$

B. $d = 3, g = 2$

C. $d = 3, g = 7$

D. $d = 5, g = 10$

6 Which of the following are possible values for b and c in the equation below if b is a negative number greater than -5?

$$c = 3 - b$$

A. $b = -2, c = 5$

B. $b = 0, c = 3$

C. $b = -7, c = 10$

D. $b = 5, c = -2$

7 What is a possible coordinate pair in the form (x, y) for the equation below if x is an odd whole number less than 5?

$$y = x + 3$$

```



```

8 Which of the following are possible coordinate pairs in the form (a, x) for the equation below if a is an even whole number less than 12?

$$a + 7 = x$$

A. (2, 8)

B. (6, 12)

C. (9, 16)

D. (10, 17)

9 Which of the following statements is true for the equation below as the value of d increases?

$$p = \frac{50}{d} - 25$$

A. p decreases.

B. p increases.

C. p stays the same.

D. p always gets closer to zero.

10 Which of the following statements is true for the equation below if the value of x is positive?

$$x = \frac{y}{10} - 50$$

A. The value of y is 0.

B. The value of y is any integer.

C. The value of y is less than x.

D. The value of y is greater than x.

11 The formula for the volume of a rectangular prism is $V = lwh$, where V is the volume, l is the length, w is the width, and h is the height. If all the dimensions of the prism are tripled, what effect does this have on the volume?

A. The volume is doubled.

B. The volume is tripled.

C. The volume is multiplied by 9.

D. The volume is multiplied by 27.

12 The formula for the area of a square is $A = s^2$, where A is the area and s is the length of a side. If the side of a square doubles, what effect does this have on the area?

A. The area is 2 times greater.

B. The area is 4 times greater.

C. The area is 8 times greater.

D. The area is 16 times greater.

MILE 21: SETTING UP EQUATIONS AND INEQUALITIES

Information on the MCAS can be provided in forms that resemble algebraic problems. For example, "I added $2.50 to my piggy bank and I now have $6.00," "John has more than twice as many toy cars as Alice has," and "This month's 13 inches of rain was half the amount of rain that fell last month" are all algebraic problems. To translate verbal expressions into algebraic expressions, you must know the mathematical meaning of certain words. Below is a list of some important definitions.

- A **variable** is a symbol (often a letter such as a) used to represent a value.

- A **coefficient** is a number multiplied by the variable, such as the 3 in $3a$.

- A **term** is a variable, a number, or a product of both, such as $3a$.

- An **expression** is a combination of variables or numbers related by at least one operation, such as $3a + 5$ or 2×3.

- An **equation** is an expression with an equal sign, such as $3a = 6$.

- An **inequality** is an expression with a sign other than an equal sign, such as $a > 5$.

If there is an unknown quantity in the sentence, you need a way to represent it. Letters represent a variable or an unknown quantity. You can usually choose which letter you want to represent as the variable. It is useful, however, to use a letter that makes sense. If the expression refers to animals, you might want to use a as the variable because a is the first letter in the word *animals*.

The following table lists some verbal expressions and their equivalent in algebraic terms. Read through the table carefully, as some of the expressions may seem very similar. The table uses x and y as the two sample variables.

Verbal Expression	Algebraic Expression
The sum of two numbers x and y	$x + y$
The difference of two numbers x and y	$x - y$
The product of two numbers x and y	$x \cdot y$ also, xy
The quotient of two numbers x and y	$x \div y$ also, $\frac{x}{y}$
The sum of x and y is fifteen.	$x + y = 15$
Eight more than x	$8 + x$
Twelve less than four times the number x	$4x - 12$
Twice the sum of x and four	$2(x + 4)$
The sum of twice x and nine	$2x + 9$
x is greater than y.	$x > y$

Notice how certain common expressions can change to math operations.

"I added $2.50" is the same as + $2.50.

"I added $2.50 to my piggy bank" is the same as p + $2.50.

"I added $2.50 to my piggy bank and I now have $6.00" is the same as p + $2.50 = $6.00.

"Half the amount of rain that fell last month" is the same as $\frac{m}{2}$.

"13 is half the amount of rain that fell last month" is the same as $13 = \frac{m}{2}$.

"Twice as many toy cars as Alice has" is the same as $2a$.

"John has more than twice as many toy cars as Alice has" is the same as $j > 2a$.

"John has no more than twice as many toy cars as Alice has" is the same as $j \leq 2a$.

Directions: Write out algebraic expressions that represent the information in each of the following sentences.

1. Jack added 10 CDs to his collection and now owns 115.

    _____

2. Sally lost 5 paper clips and now has fewer than 50.

    _____

3. The product of Theresa and Alonzo's home run totals is equal to Sammy's home run total.

    _____

4. Jill received a score on her science test that was 30 points higher than the quotient of Miguel and Cathy's scores.

    _____

5. Two times Lynn's age minus 7 is greater than or equal to 14.

    _____

Directions: Read each question carefully. Circle your answer choice for each question.

1. Lennox made $150 baby-sitting last month. He wants to buy a few CDs and save at least $30. If each CD costs $15, which inequality can be used to find the number of CDs, c, that Lennox can buy?

   A. $150 - \$30c \le \$15$

   B. $150 - \$30c \ge \$15$

   C. $150 - \$15c \le \$30$

   D. $150 - \$15c \ge \$30$

2. Kimberly practices soccer each day. She made a chart to show the number of shots she made and the number of goals she scored.

Shots (s)	Goals (g)
57	19
39	13
84	28
27	9

   Which of the following equations shows the relationship between s, the number of shots Kimberly made, and g, the number of goals she scored?

   A. $g + 42 = s$

   B. $g - 42 = s$

   C. $3g = s$

   D. $\frac{1}{3}g = s$

3. Ola stocked a shelf in a store with bottles of juice. She had 44 bottles to place on four shelves. She placed the same number of bottles on each shelf and had 8 bottles left over. Which of the following equations can be used to find the number of bottles, b, on each shelf?

   A. $4b + 8 = 44$

   B. $4b - 8 = 44$

   C. $\frac{b}{4} + 8 = 44$

   D. $\frac{b}{4} - 8 = 44$

4. Julia figures that it will cost between $30 and $50 to attend a concert. She earns $5 for each lawn she mows. Which of the following shows the number of lawns, l, Julia will need to mow to make enough to go to the concert?

   A. $\$30 \ge 5l \ge \$50$

   B. $\$30 \le 5l \le \$50$

   C. $\$30 \le \frac{5}{l} \le \$50$

   D. $\$30 \ge \frac{5}{l} \le \$50$

5. Dorian is 4 and his brother Marquis is 12. Which equation shows the relationship between Dorian's age, d, and Marquis's age, m, every year?

   A. $m - d = 8$

   B. $m + d = 12$

   C. $3d = m$

   D. $3m = d$

6 A computer printer is selling for $245. The printer and the cables together will not cost more than $300. Which of the following inequalities shows the possible cost of the cables, *c*?

A. $\$245 + c \le \$300$

B. $\$245 + \$300 \le c$

C. $\$300 - \$245 \le c$

D. $\$300 + c \le \$245$

7 Cameron bought a roll of film that takes 36 pictures. He wants to take at least 20 pictures at his graduation. Which expression can be used to find the number of pictures, *p*, that Cameron can take before his graduation?

A. $36 - 20 = p$

B. $36 - 20 \le p$

C. $36 - p \le 20$

D. $36 - p \ge 20$

8 Zamir has 31 math textbooks to pack into boxes. He puts 7 books in each box and has 3 books left over. Which of the following equations can be used to find the number of boxes, *b*, Zamir used?

A. $7b - 3 = 31$

B. $7b + 3 = 31$

C. $\frac{b}{4} - 3 = 31$

D. $\frac{b}{7} + 3 = 31$

9 At a research farm, the height of a seedling was recorded each week as shown in the chart below.

Week (*w*)	Height (*h*)
1	3
2	5
3	6
4	9

Which of the following equations approximates the relationship between *w*, the week, and *h*, the height of the seedling?

A. $w + 2 = h$

B. $w - 2 = h$

C. $2w = h$

D. $\frac{1}{2}w = h$

10 Ms. Yan has 56 tomato plants. She plants 6 in each row and has 2 left over. Which of the following equations can be used to find the number of rows, *r*, Ms. Yan planted?

A. $56 = 6r + 2$

B. $56 = 6r \div 2$

C. $\frac{56}{6} + 2 = r$

D. $\frac{56}{6} - 2 = r$

11 A shrub grows 2 inches each year, while a neighboring tree grows 6 inches each year. Which equation shows the relationship between the growth of the shrub, *s*, and the growth of the tree, *t*, every year?

A. $t - s = 4$

B. $t + s = 8$

C. $3t = s$

D. $3s = t$

MILE 22: SOLVING ALGEBRAIC EQUATIONS

Algebraic expressions can be simplified and solved using the order of operations. Whenever you are given several mathematical operations to perform in one problem, use the Order of Operations. That's PEMDAS, which can be more easily remembered as the sentence "Please Excuse My Dear Aunt Sally." Below you will find the meaning of each initial.

In any algebraic expression,

P: Solve any math that is placed within the parentheses first.

E: Solve any exponents in the expression.

M/D: Solve any multiplication and division, in order from left to right.

A/S: Solve any addition and subtraction, in order from left to right.

Example 1: $2^2 \times (5 - 3) + 4 \times 3 \div 6 = x$

$2^2 \times (2) + 4 \times 3 \div 6 = x$

$4 \times 2 + 4 \times 3 \div 6 = x$

$8 + 4 \times 3 \div 6 = x$

$8 + 12 \div 6 = x$

$8 + 2 = x$

$10 = x$

Algebra problems will usually ask you to solve for the unknown value. In this example, the variable x was already by itself on one side of the equal sign. All you had to do was simplify the terms on the left side of the equation to find that $x = 10$. Sometimes you will need to manipulate the numbers to get the variable by itself on one side of the equal sign.

Example 2: $4m + 15 = 135$

When you try to solve for this variable, work from the terms farthest from the variable. Notice that you are adding 15 to the term $4m$ in the example above. Undo the addition by subtracting 15 from both sides of the equation to get $4m = 120$. Then undo the multiplication by dividing each side of the equation by 4 to get $m = 30$.

Two-step solutions can be applied to real-life situations. Look at the example below. This situation uses the same equation from Example 2 to find the solution.

Example 3:   **Henri needs to withdraw equal amounts of money from his 4 piggy banks and add this money to the $15 he already has so that he can pay for a $135 radio. How much money was taken from each piggy bank?**

The variable m represents the amount of money from each piggy bank. You would write the equation $4m + \$15 = \$135$ and solve it like you did Example 2.

Multistep algebraic expressions may involve the use of several mathematical operations.

Example 4: 2(4*m* + 15) = 270

In this case, you wouldn't undo the addition first, because it is inside the parentheses. You could choose to use the distributive property to change the equation to 8*m* + 30 = 270. Then you can solve by undoing the addition and multiplication, like in Example 2.

You could also choose to solve the equation by undoing the outermost operation, multiplying by 2. Undo the multiplication by dividing both sides of the equation by 2 to get 4*m* + 15 = 135. Then you can solve by undoing the addition and multiplication, like in Example 2.

Directions: Fill in the following puzzle using the clues given below.

Across:

1. If $x = 3$, what is $2x + 15$?

2. If $z = 5$, what is $20z + \frac{8z}{2} + 22$?

4. What is the largest integer y can be if $y < 9 \times 10(3 + 7) + 3$?

Down:

1. If $y = 20$, what is $5^2 + 9y - 1$?

2. If $m = 12$, what is $12m + \frac{3m}{2} + 7$?

3. What is the smallest integer n can be if $n > 20 \times (27 - 17) + 11$?

Directions: Read each question carefully. Write your answer in the space below the question or circle the letter next to your answer choice.

1   What is the value of *n* in the following equation?

$$5n - 3(n + 4) = -6$$

A.   −3

B.   3

C.   5

D.   12

2   Solve for *y* in the equation below.

$$4(y - 2) + 8y = 88$$

A.   4

B.   6

C.   8

D.   12

3   In the equation $8n - 3(2n + 4) = 6$, what is the value of *n*?

4   In the equation $5x + 2(2x - 4) = 19$, what is the value of *x*?

5   In the equation $3(2t + 4) - 8t = 14$, what is the value of *t*?

6   What is the value of *h* in the following equation?

$$8 + \frac{(6h)}{4} = 14$$

7 What is p in the equation $\frac{(2p)}{6} - 4 = 1$?

A. 9

B. 15

C. 25

D. 30

8 Which of the following would lead to the correct solution for the equation below?

$$16 - 5y = 1$$

A. Subtract 16 from both sides, then divide by -5.

B. Add 16 to both sides, then divide by 5.

C. Subtract 16 from both sides, then multiply by 5.

D. Add 16 to both sides, then multiply by -5.

If you get stuck trying to solve an equation for a multiple-choice question, replace the variable with the value in each answer choice until you find the correct one.

9 Which of the following would lead to the correct solution for the equation below?

$$3m - 4 = 8$$

A. Subtract 4 from both sides, then divide by -3.

B. Add 4 to both sides, then divide by 3.

C. Subtract 4 from both sides, then multiply by 3.

D. Add 4 to both sides, then multiply by 3.

10 Which of the following would solve the following equation?

$$25 - 4a = 5$$

A. Subtract 25 from both sides, then divide by -4.

B. Add 25 to both sides, then divide by 4.

C. Subtract 25 from both sides, then multiply by 4.

D. Add 25 to both sides, then multiply by -4.

11 Which of the following would solve the following equation?

$$\frac{x}{4} + 2 = 6$$

A. Add 2 to both sides, then multiply by -4.

B. Add 2 to both sides, then divide by 4.

C. Subtract 2 from both sides, then multiply by -4.

D. Subtract 2 from both sides, then multiply by 4.

MILE 23: FUNCTIONS

In mathematics a **function** is a relationship between an input and an output, based on a certain formula. To find that relationship you just need to identify the rule for each situation. Look at a few examples.

Example 1: **If Janie makes $5 for every necklace sold, what function shows how much money she could make?**

There are two unknowns in this situation, so use two variables: n for the number of necklaces sold and m for the amount of money Janie makes. If Janie makes $5 for every necklace sold, she makes $5 for selling 1 necklace and $10 for selling 2 necklaces. For every different n, the number of necklaces sold, Janie makes a different output m. The different values of n and m can be represented in a function table like the one below. You can see from the table that m is always 5 times greater than n, so the function is written $5n = m$.

Input (n)	0	1	2	3
Output (m)	0	5	10	15

Example 2: **Arnie wants to know how much money he will have remaining after he buys a present for a number of his friends. Each present will cost $6, and Arnie begins shopping with $62. What function table demonstrates this situation?**

The first step is to identify the numbers that can change in the problem. You know that if Arnie buys one present, he will have $62 − $6(1) = $56 remaining. If he buys two presents, he will have $62 − $6(2) = $50 remaining. The two amounts that change are the number of presents, p, and the money remaining, r. The equation that shows the functional relationship between the variables is $62 − 6p = r$. Plug in different values of p to come up with the function table representing this relationship.

No. of presents	Expression	$ Remaining
1	$62 − 6(1) = r$	56
2	$62 − 6(2) = r$	50
3	$62 − 6(3) = r$	44
4	$62 − 6(4) = r$	38

Sometimes you may be given a table and asked to find the rule that relates the input and the output values. This could be a pretty difficult question to answer. Fortunately, you will usually be given a few answer choices, one of which must be correct. Insert the x and y values into each function and see which one works for all the values.

Work through an example of finding the rule without any answer choices. In this example a function will find the number of diagonals in different polygons. Look at the polygons below. Count the number of sides on each polygon and fill in the *n* row in the table below.

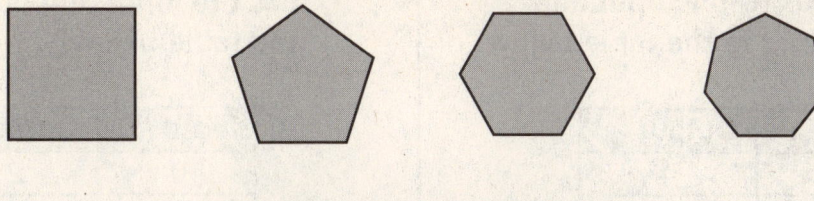

Number of Sides (*n*)	Number of Diagnals (*d*)

Now draw and count the number of diagonals in each polygon. Remember that a diagonal is a line segment that connects two corners or vertices of the polygon, but it doesn't form a side of the polygon. You should find that a quadrilateral has 2 diagonals. You should also find that a pentagon has 5 diagonals and a hexagon has 9 diagonals. Fill in those numbers in the table. It gets more complicated to find the number of diagonals for polygons with more sides, so let's stop right here and start looking for a rule.

Think about what you know.

- The number of sides of a polygon, *n*, equals the number of vertices, and diagonals connect vertices.

- Looking at one vertex in each of the polygons above, each vertex forms only 1 diagonal in the 4-sided shape, 2 diagonals in the 5-sided shape, and 3 diagonals in the 6-sided shape. The number of diagonals that each vertex can form is 3 less than the number of sides in the shape, $(n - 3)$. There are *n* vertices in each shape, so the number of diagonals must be $n(n - 3)$, right?

- Check the numbers in the table. If $n = 4$, then $d = 4(4 - 3) = 4$. This doesn't look right; the table says that there should be 2 diagonals for a quadrilateral. If $n = 5$, then according to the rule, $d = 5(5 - 3) = 10$. Uh-oh, it looks like the rule is counting each diagonal twice.

- Divide the rule by 2 to get $d = \frac{n(n - 3)}{2}$ and check it with the values in the table. If $n = 4$, then $d = \frac{4(4 - 3)}{2} = 2$. If $n = 5$, then $d = \frac{5(5 - 3)}{2} = 5$. It works!

Directions: Use the function $d = \frac{n(n - 3)}{2}$ to finish filling in the table above.

Directions: Read each question carefully. Circle your answer choice for each question.

1   Which of the following equations corresponds to the input and output values in the table below?

Input (x)	Output (y)
0	4
1	3
2	4
3	7

A. $y = x^2 - 2x + 4$

B. $y = x^2 - 2x$

C. $y = x^2 + 2x + 4$

D. $y = x^2 - 4$

2   Which of the following equations corresponds to the input and output values in the table below?

Input (x)	Output (y)
0	−1
1	1
2	5
3	11

A. $y = x^2 - 1$

B. $y = x^2 + x - 1$

C. $y = x^2 + 2x + 1$

D. $y = x^2 - 4x$

3   Which of the following equations has the input and output values in the table below?

Input (x)	Output (y)
0	2
1	4
2	6
3	8

A. $y = x - 2x$

B. $y = x + 2x$

C. $y = 2x + 2$

D. $y = 4x + 4$

4   Which of the following equations has the input and output values in the table below?

Input (x)	Output (y)
0	−8
1	−7
2	−4
3	−1

A. $y = x^2 - 8$

B. $y = x + 2$

C. $y = 2x + 2$

D. $y = 4x + 2$

5 Which of the following equations has the input and output values in the table below?

Input (x)	Output (y)
0	2
2	6
4	10
6	14

A. $2(a - 2)$

B. $2a + 2$

C. $a + 2$

D. $4a + 2$

6 Which of the following equations has the input and output values in the table below?

Input	Output
1	5
7	41
8	47
9	53
r	?

A. $r + 4$

B. $r - 6$

C. $4(r + 1)$

D. $6r - 1$

7 Which of the following equations has the input and output values in the table below?

Input	Output
0	−2
2	2
4	6
6	10
p	?

A. $2p - 2$

B. $2(p + 2)$

C. $4p + 4$

D. $4p + 8$

8 Which of the following equations has the input and output values in the table below?

Input (x)	Output (y)
−4	8
1	13
5	17
10	22
t	?

A. $t - 4$

B. $t + 12$

C. $2t + 4$

D. $4t - 2$

MILE 24: GRAPHING LINEAR EQUATIONS

A **linear equation** is a function with two variables, x and y. The corresponding values of x and y that fit into the function form ordered pairs (x, y) that can be graphed on a coordinate grid. The number you get for x will correspond to the x-axis, and the number you get for y will correspond to the y-axis. For linear equations, the graph of the equation will form a straight line.

An example of a linear equation is $y = 2x + 1$.

You can graph this linear equation. The table below shows some numbers that work in the equation. The numbers give you ordered pairs that you can plot on a graph.

x	Expression	y	(x, y)
−2	2(−2) + 1	−3	(−2,−3)
−1	2(−1) + 1	−1	(−1,−1)
0	2(0) + 1	1	(0, 1)
1	2(1) + 1	3	(1, 3)
2	2(2) + 1	5	(2, 5)

Linear equations allow you to choose any value for x and solve for y. In the chart above, the values for x were -2, -1, 0, 1, and 2. For each x-value there is a y-value. Together they create an ordered pair to plot on the graph.

Using the ordered pairs in the table above, graph the linear equation $y = 2x + 1$.

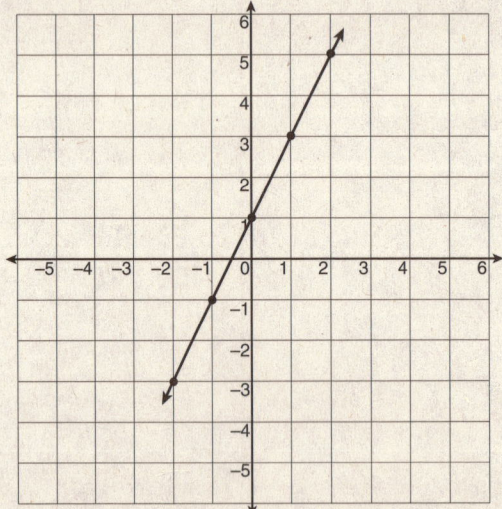

Linear equations are written in the form $y = mx + b$, where m is the slope and b is the y-intercept. Slope describes the "steepness" of a line. In this example, 2 is the slope. Find the slope by taking two points on the line and calculating the change in y over the change in x. The y-intercept describes where the line crosses the y-axis. In this example, the y-intercept is 1.

Directions: You have a new computer game in which you are a monkey traveling through the trees in a jungle. You face many different obstacles as you move across the screen. You start at point (–2, –7) and move according to the equation $y = 2x - 3$. Chart your progress by figuring out at least four solutions to the equation that will give you ordered pairs to plot on the grid. Plot at least four more points on the grid below and then connect the points with a line.

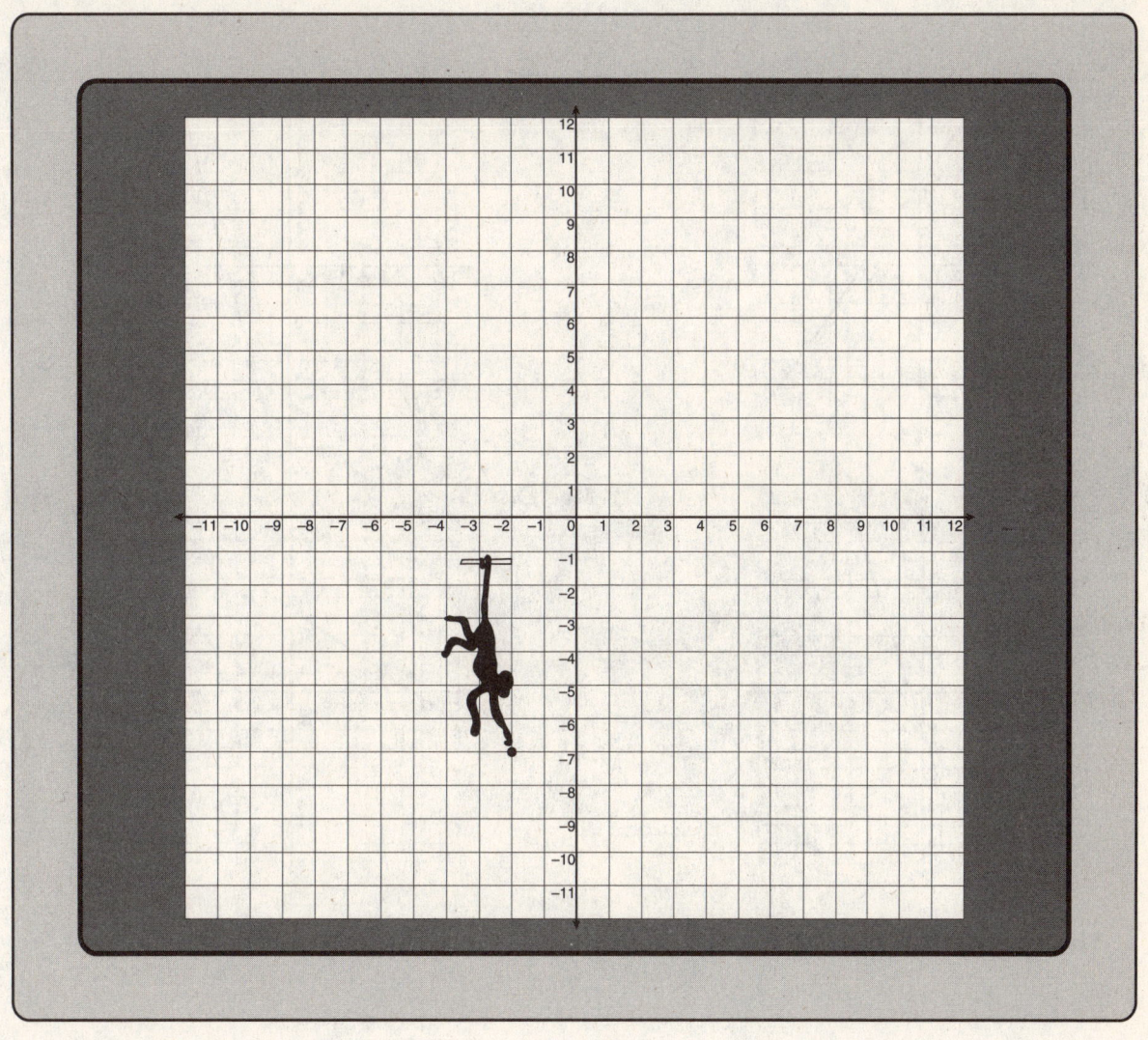

Directions: Read each question carefully. Circle the correct answer choice for each question.

1   Which of the lines plotted below is the correct representation for the pairs of points listed below?

$$(-2, -8), (-1, -5), (2, 4)$$

A

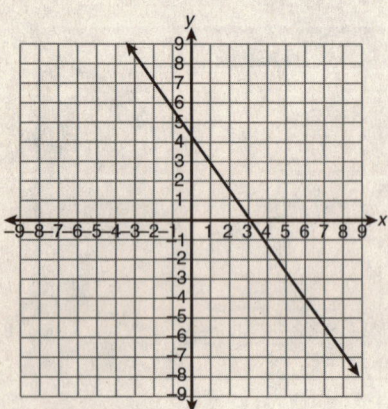

C

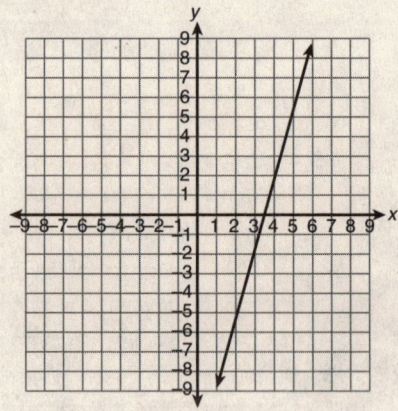

B

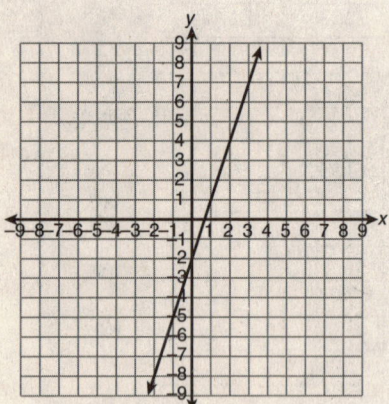

D

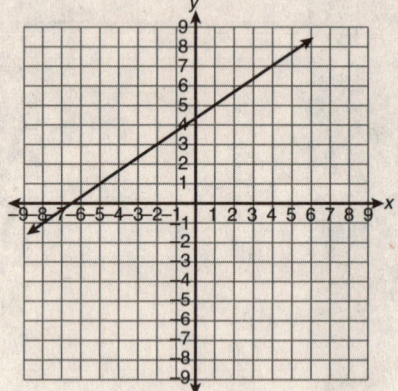

2   The graph of which of the following equations includes the point (2, 11)?

A. $y = 4x - 3$

B. $y = 3x - 3$

C. $y = 4x + 3$

D. $y = 3x + 3$

Remember that the first number in an ordered pair corresponds to the x-value and the second number corresponds to the y-value. It is important to keep this in mind as you graph linear equations.

3 Which points are on the following line?

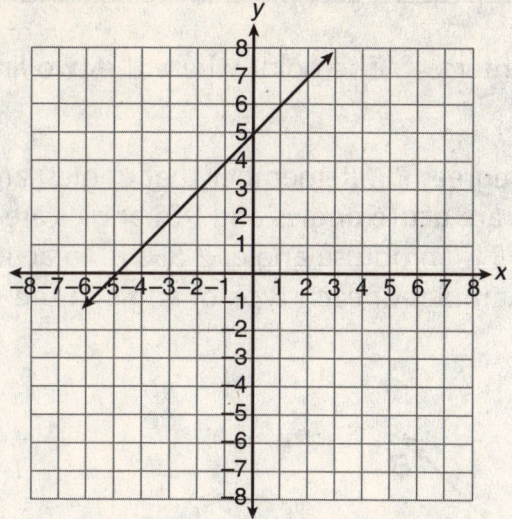

A. (−2, 3) and (0, 5)

B. (−2, 5) and (0, −3)

C. (2, 5) and (0, 3)

D. (2, 6) and (0, 5)

4 Which points are on the following line?

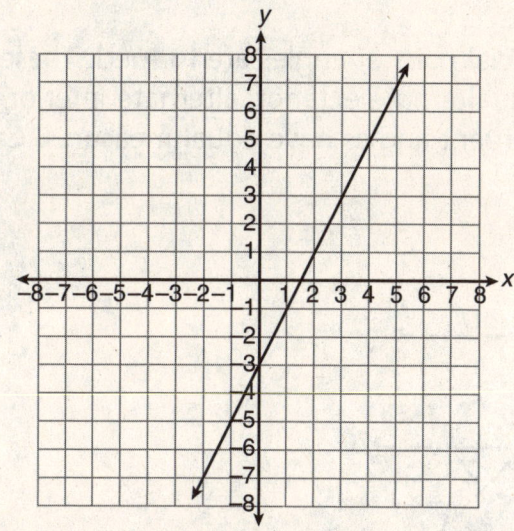

A. (−1, −5) and (0, 1)

B. (−1, −5) and (2, 1)

C. (1, 5) and (0, 1)

D. (1, 5) and (2, 1)

5 The graph of the linear equation $y = 2x − 4$ should include which of the following points?

A. (0, −3)

B. (−3, 7)

C. (2, 1)

D. (1, −2)

6 The graph of the linear equation $y = 3x + 2$ should include which of the following points?

A. (0, 3)

B. (1, 5)

C. (2, −4)

D. (3, 10)

7 The graph of the linear equation $y = −2x + 7$ should include which of the following points?

A. (−1, 9)

B. (0, −2)

C. (2, 7)

D. (4, 15)

MILE 25: ANGLES

An **angle** is a figure that is formed when two lines intersect. The point where the two lines meet is called the **vertex.**

Angles can be classified by their measurements in degrees (°). Remember that angles larger than 90° are **obtuse** angles angles smaller than 90° are **acute** angles and 90° angles are **right** angles. For example, in the figure below, ∠AEB is an obtuse angle, ∠BEC is an acute angle, and ∠FEC is a right angle. There are more examples of each type of angle in the figure below. Can you find them?

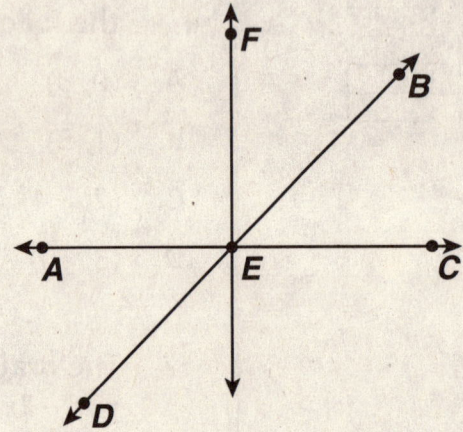

There are also special pairs of angles in the diagram above. **Complementary angles,** ∠FEB and ∠CEB, are two angles with a sum of 90°. **Supplementary angles,** ∠AED and ∠CED, are two angles with a sum of 180°.

When a line crosses a pair of parallel lines, more special pairs of angles are formed. These pairs are listed in the box below. Notice that vertical, alternate exterior, alternate interior, and corresponding angle pairs are congruent. Congruent angles have equal measures. Can you also find supplementary angles in this diagram?

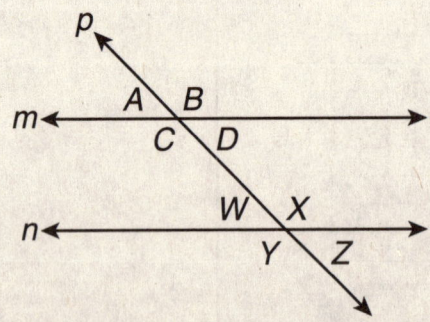

Vertical angles	∠A and ∠D; ∠B and ∠C; ∠W and ∠Z; ∠X and ∠Y
Alternate exterior angles	∠A and ∠Z; ∠B and ∠Y
Alternate interior angles	∠C and ∠X; ∠D and ∠W
Corresponding angles	∠A and ∠W; ∠B and ∠X; ∠C and ∠Y; ∠D and ∠Z

Directions: Read each question carefully. Circle the correct answer choice for each question.

1   Line *TU* is parallel to line *VW*. What is the measure of ∠*AYW*?

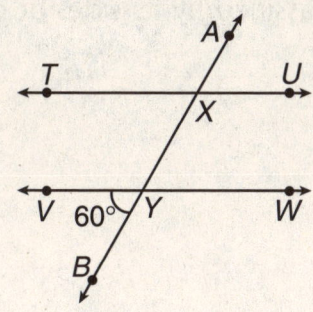

A.   30°

B.   60°

C.   90°

D.   120°

2   What is the measure of ∠*MXP* in the diagram below?

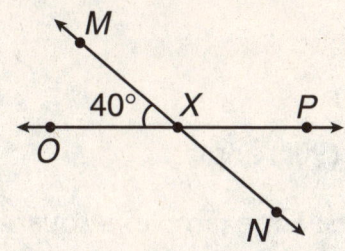

A.   40°

B.   50°

C.   100°

D.   140°

3   A pie is cut into 8 equal slices. Which of the following is the best estimate for the number of degrees of one slice of the pie?

A.   30°

B.   45°

C.   90°

D.   120°

4   First Street runs parallel to Second Street. A railroad track intersects both streets, as shown in the diagram.

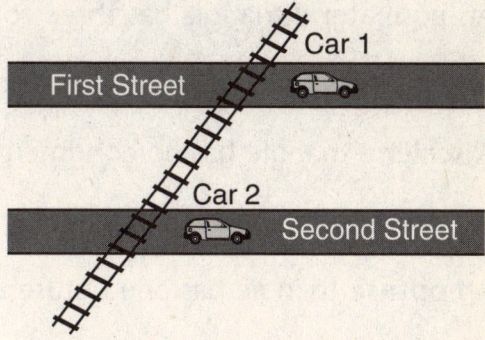

If First Street and the railroad tracks make a 60° angle at Car 1, what angle is formed by Second Street and the railroad tracks at Car 2?

A.   60°

B.   120°

C.   180°

D.   360°

5   A circle is cut into 10 equal parts. Which of the following is the best estimate for the number of degrees of one part of the circle?

A.   36°

B.   90°

C.   135°

D.   180°

MILE 26: PROPERTIES OF POLYGONS

A **polygon** is a closed figure made up of at least three line segments called sides. Polygons that have only three sides are **triangles.** A triangle can be classified by its sides or its angles.

- An **equilateral** triangle has three congruent sides.

- An **isosceles** triangle has two congruent sides.

- A **scalene** triangle has no congruent sides.

- An **acute** triangle has three acute angles.

- An **obtuse** triangle has one obtuse angle.

- A **right** triangle has one right angle.

Polygons that have four sides, such as squares, are called **quadrilaterals.** There are several types of quadrilaterals. Each one has certain properties.

- A **trapezoid** has one pair of parallel sides.

- A **parallelogram** has both pairs of opposite sides parallel.

- A **rhombus** has four sides of equal length.

- A **rectangle** has four right (90°) angles.

- A **square** has four sides of equal length and four right (90°) angles.

Notice how some shapes are actually types of other shapes. For example, a square is actually a type of rectangle, as well as a type of rhombus, and a rhombus is a type of a parallelogram.

Polygons can have any number of sides, and there is a name for each polygon with a different number of sides. The names of polygons are based on the prefix for that number.

penta = 5   pentagon = 5-sided polygon   hexa = 6   hexagon = 6-sided polygon

hepta = 7   heptagon = 7-sided polygon   octa = 8   octagon = 8-sided polygon

nona = 9   nonagon = 9-sided polygon   deca = 10   decagon = 10-sided polygon

Another property of polygons is the sum of their exterior and interior angles. The sum of the exterior angles always remains the same, 360°. Some of the sums of the interior angles are as follows:

triangle = 180°   pentagon = 540°   octagon = 1,080°

quadrilateral = 360°   hexagon = 720°   decagon = 1,440°

A polygon whose sides and angles are congruent is a **regular** polygon.

Directions: Write the word that best fits the description of each shape below. Then find those words in the word search puzzle below.

1.   A four-sided polygon. _____

2.   A polygon whose interior angles measure 540°. _____

3.   A triangle with a right angle. _____

4.   A planar figure with many sides. _____

5.   A four-sided figure with only one pair of parallel sides. _____

6. A closed figure with nine sides. _____

7.   A four-sided figure with no right angles and two pairs of parallel sides. All sides are congruent. _____

8.   A triangle with three congruent sides. _____

```
C N W Q A B E P M A T H B M
A D O U T L D E N F H A B U
B F G P R E C N V K G L J K
E L M O H C G T F I I Q W Z
Q U A D R I L A T E R A L E
U Z A N O G I G Y G H C F J
I X B D E B P O L Y G O N D
L N A P Q H W N O N A G O N
A P O Y I J R R S U D C P Q
T R A P E Z O I D T S T L M
E Y F K U M S P E X Q U O N
R H O M B U S V B J K V T A
A Z G L M N T Q X O R L N P
L H I R N S W P Z Y P A C E
```

Directions: Read each question carefully. Write your answer on a separate sheet of paper or circle the letter next to your answer choice.

1   Mitzi drew some regular polygons and divided each into triangles. She knew that the sum of the interior angles of a triangle is 180°. Mitzi made the chart below.

Polygon	Number of Triangles	Sum of Interior Angles
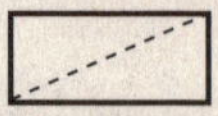	2	360°
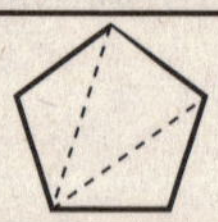	3	540°
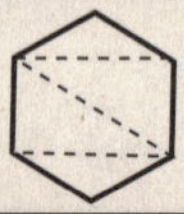	4	720°
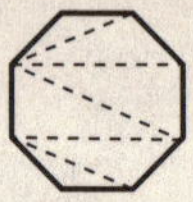	6	1080°

a. What is the relationship between the number of sides of a polygon and the number of triangles it can be divided into?

b. A polygon has 9 sides. What is the sum of the interior angles of the polygon?

c. Describe how you can use the number of sides of any polygon to find the sum of the interior angles of the polygon.

d. If the sum of the interior angles of a polygon is 1,800°, how many sides does the polygon have?

2 **Which of the following measurements will not form a triangle?**

A. 6 in., 3 in., 6 in.

B. 6 in., 6 in., 6 in.

C. 6 in., 8 in., 10 in.

D. 6 in., 9 in., 18 in.

3 **Which of the following measurements will not form a triangle?**

A. 4 in., 3 in., 4 in.

B. 8 in., 8 in., 8 in.

C. 2 in., 3 in., 5 in.

D. 6 in., 9 in., 5 in.

4 **Which of the following is not always a true statement about polygons?**

A. A rhombus is a parallelogram.

B. A square is a rectangle.

C. A trapezoid is a quadrilateral.

D. A parallelogram is a rectangle.

5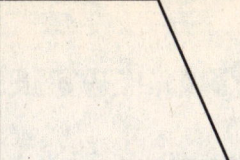

Which of the following is always true of a trapezoid?

A. It has a right angle.

B. It is a regular polygon.

C. It has one pair of parallel sides.

D. It has at least one pair of congruent angles.

6 **Kamal described a figure whose interior angles are all congruent and add up to 720°. The sides of the figure are also congruent. Which polygon did Kamal describe?**

A. a regular hexagon

B. an irregular hexagon

C. a regular octagon

D. an irregular octagon

MILE 27: PROPERTIES OF SOLIDS

Solids are three-dimensional figures made up of two-dimensional sides called faces. You'll recognize these faces as familiar polygons. The line where two faces join is called an **edge**. The point where edges intersect is called a vertex or plural, vertices.

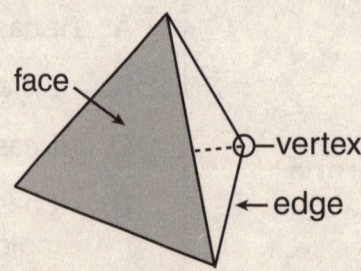

Prisms are solids with identical polygons on the top and bottom and parallelograms for sides. The polygons on the top and bottom are called bases. A prism is identified by the shape of its base. Below are examples of prisms.

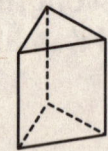 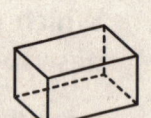 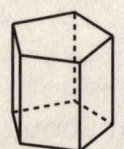 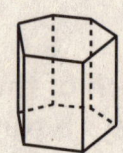

Triangular Prism **Rectangular Prism** **Pentagonal Prism** **Hexagonal Prism**

Pyramids are solids with a polygon base and triangles for sides. Pyramids are identified by the shape of their bases. Below are examples of pyramids.

 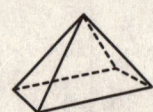 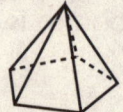

Triangular Pyramid **Rectangular Pyramid** **Pentagonal Pyramid**

Some solids have circular faces. A **cylinder** resembles a prism but has circles for bases. A **cone** resembles a pyramid but has a circle for a base.

 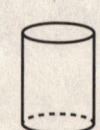

Cone **Cylinder**

Nets are figures that result when a solid is cut at the edges and spread out flat. Below are examples of nets.

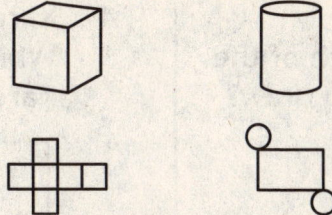

Directions: Look at the three-dimensional objects below. To the right of the objects are six shapes. Draw a line from each three-dimensional object to the shape that shows the net of the object. The nets are not to scale.

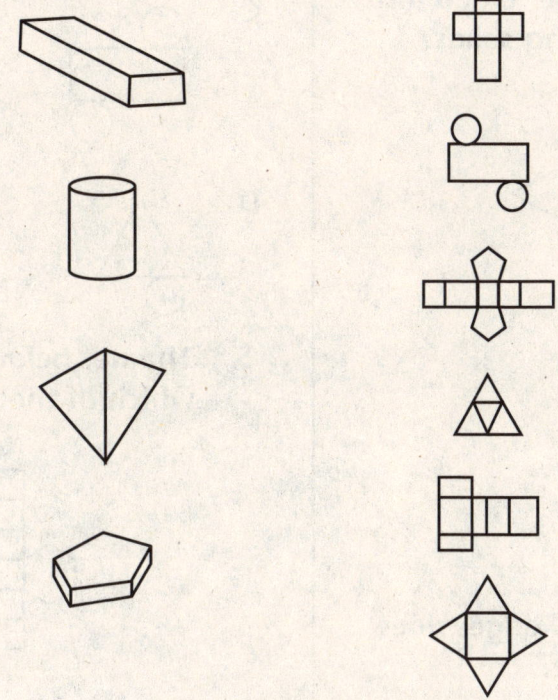

Directions: Read each question carefully. Circle the correct answer choice for each question.

1 Nikko drew the net of a rectangular pyramid. Which of the following shapes did Nikko draw?

A. 5 rectangles

B. 5 triangles

C. 4 rectangles and 1 triangle

D. 1 rectangle and 4 triangles

2 The net below can be folded into which of the following solids?

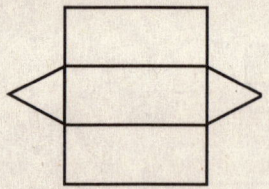

A. rectangular prism

B. rectangular pyramid

C. triangular prism

D. triangular pyramid

3 How many faces and edges does the solid below have?

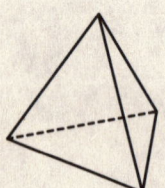

A. 3 faces, 6 edges

B. 4 faces, 6 edges

C. 5 faces, 9 edges

D. 6 faces, 12 edges

4 Karen drew the net of a cylinder. Which of the following shapes did Karen draw?

A.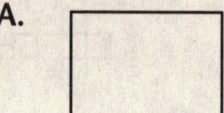

B.

C.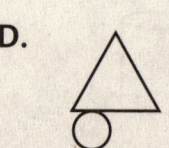

D.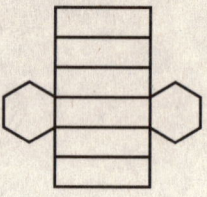

5 The net below can be folded into which of the following solids?

A. hexagonal prism

B. rectangular pyramid

C. rectangular prism

D. hexagonal pyramid

6   Vanessa made the chart below to show the number of vertices, faces, and edges in solid figures.

Shape	Number of Faces	Number of Verticies	Number of Edges
Triangular Prism	5	6	9
Rectangular Prism	6	8	12
Pentagonal Prism	7	10	15
Hexagonal Prism	8	12	18
Triangular Pyramid	4	4	6
Rectangular Pyramid	5	5	8
Pentagonal Pyramid	6	6	10
Hexagonal Pyramid	7	7	12

a. How many faces, vertices, and edges would an octagonal prism have?

b. What is the relationship between the number of faces, f, in a prism and the number of vertices, v?

c. How many faces, vertices, and edges would an octagonal pyramid have?

d. What is the relationship between the number of faces in a pyramid, f, and the number of edges, e?

MILE 28: CONGRUENCE AND SIMILARITY

When you look around you may notice that shapes repeat themselves. Paving stones, street signs, and patterns on wallpaper are repeating shapes. You can describe repeating shapes as congruent or similar. When two figures are **congruent**, they have the exact same size and shape.

In pairs of congruent figures, parts that match are called **corresponding parts.** Check out the two triangles shown below.

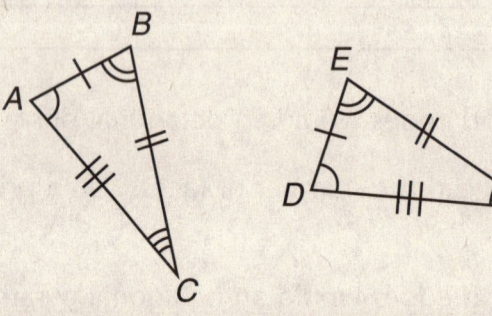

△*ABC* is congruent to △*DEF.* The symbol ≅ means "is congruent to." Notice that all corresponding parts of congruent figures are also congruent.

Congruent angles	Congruent sides
$\angle A \cong \angle D$	$\overline{AB} \cong \overline{DE}$
$\angle B \cong \angle E$	$\overline{BC} \cong \overline{EF}$
$\angle C \cong \angle F$	$\overline{AC} \cong \overline{DF}$

You don't need to know that all corresponding parts of some shapes are congruent in order to prove that the shapes are congruent. There are three theorems you can use to prove that two triangles are congruent.

- **Side, Angle, Side (SAS)** If two sides and the included angle of one triangle are congruent to two sides and the included angle on another triangle, then the two triangles are congruent.

- **Angle, Side, Angle (ASA)** If two angles and the side between them are congruent in two triangles, then the triangles are congruent.

- **Side, Side, Side (SSS)** If all three sides of two different triangles are congruent, then the triangles are congruent.

When two figures are **similar,** they have the same shape but not necessarily the same size. Similar figures have corresponding angles that are congruent and corresponding sides that are in proportion. This means that the lengths of the corresponding sides in a pair of similar figures all have equal ratios. If one side of a polygon is three times as large as its corresponding side in a similar polygon, then the ratio of the lengths of all the corresponding sides is 3 to 1.

Directions: Examine the four triangles below. Use the information provided about the triangles to determine whether each statement below is true or false. Write "T" or "F" in the blank provided before each statement.

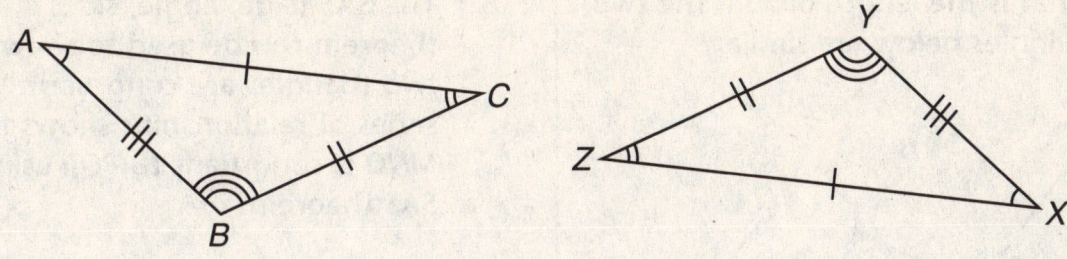

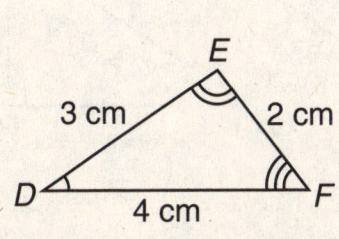

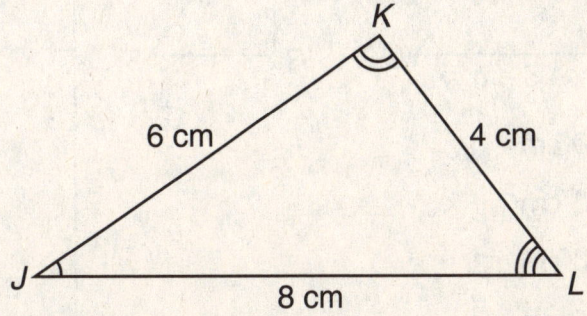

1. _____ In △ABC and △XYZ, ∠B is congruent to ∠Z.

2. _____ △DEF and △JKL are similar because the lengths of their corresponding sides have equal ratios.

3. _____ You could use the side, angle, side (SAS) theorem to prove that △ABC and △XYZ are congruent.

4. _____ The ratio between the corresponding sides in △DEF and △JKL is 3 to 1.

5. _____ In △ABC and △XYZ, ∠A is congruent to ∠X.

6. _____ △DEF and △JKL are congruent because all three of their corresponding angles are equal.

7. _____ You could use the angle, side, angle (ASA) theorem to prove that △DEF and △JKL are congruent because they both have one side that measures 4 centimeters.

8. _____ You could use the side, side, side (SSS) theorem to prove that △ABC and △XYZ are congruent.

Directions: Read each question carefully. Circle the letter for your answer choice for each question.

1 What is the length of $\overline{RS}$ if the two triangles below are similar?

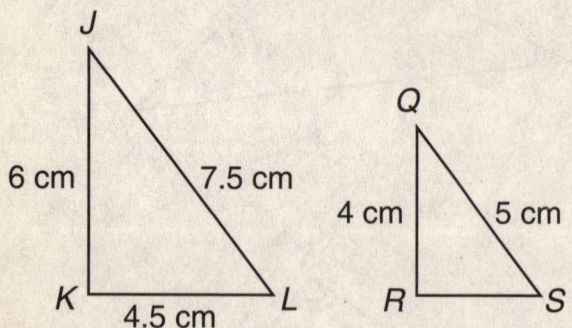

A. 4.0 cm

B. 3.0 cm

C. 2.0 cm

D. 1.5 cm

2 Evaluate the two triangles shown below.

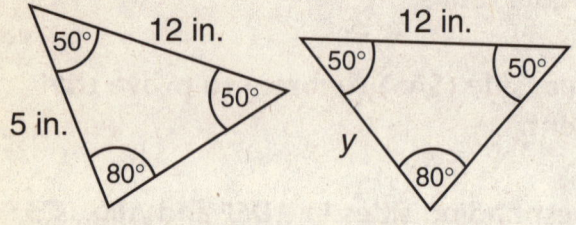

What is the length of the side labeled y?

A. 12 in.

B. 24 in.

C. 5 in.

D. 3 in.

3 The SAS (side, angle, side) theorem can be used to show that two triangles are congruent. Which series of relationships shows that *MNO* is congruent to *PQR* using the SAS theorem?

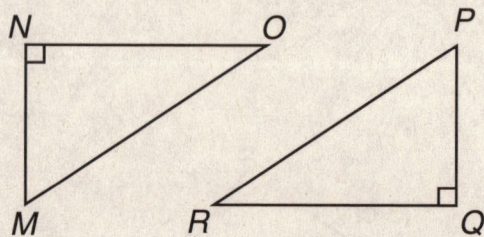

A. $\overline{NO} \cong \overline{QR}$, $\angle N \cong \angle Q$, $\overline{MN} \cong \overline{PQ}$

B. $\angle O \cong \angle Q$, $\overline{NO} \cong \overline{QR}$, $\angle O \cong \angle R$

C. $\angle O \cong \angle Q$, $\overline{NO} \cong \overline{QR}$, $\angle O \cong \angle R$

D. $\overline{NO} \cong \overline{QR}$, $\angle R \cong \angle M$, $\overline{MN} \cong \overline{RP}$

Tip

Remember that you don't always need to know that all of the corresponding angles and sides are congruent in order to figure out whether two triangles are congruent. Sometimes you can prove that the triangles are congruent if you know that some of their corresponding parts are congruent. You can use the three theorems (SAS), (ASA), and (SSS).

4 Gina needed a picture for her school yearbook. The only one she could find was too small, so she brought it to the photo store to have it enlarged. If the picture was enlarged in proportion, what could be the new dimensions of the picture?

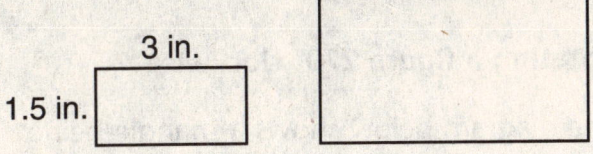

A. 4 in. by 3 in.

B. 6 in. by 3 in.

C. 6 in. by 1.5 in.

D. 9 in. by 5 in.

5 The two triangles shown below are similar to each other. What is the measurement of $\overline{BC}$?

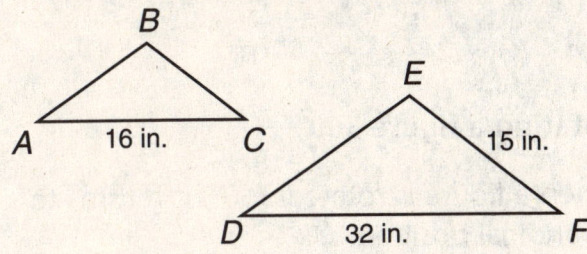

A. 30.0 in.

B. 24.5 in.

C. 8.0 in.

D. 7.5 in.

6 What is the length of $\overline{GH}$ if figure ABCD is similar to figure EFGH?

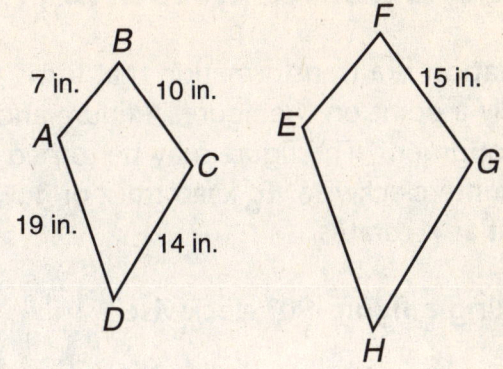

A. 21 in.

B. 19 in.

C. 17 in.

D. 16 in.

7 What is the measure of $\overline{VT}$ if $\triangle PQR$ is similar to $\triangle TUV$?

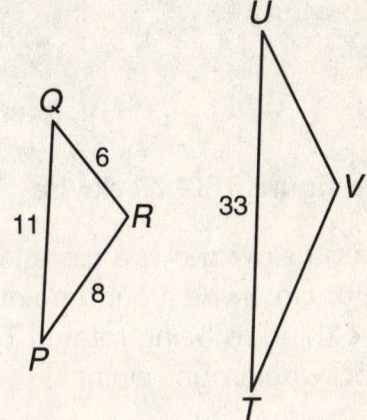

A. 16

B. 22

C. 24

D. 32

MILE 29: ROTATIONS

A **rotation** is a transformation that turns a figure about a fixed point. The fixed point is usually a point on the figure. A figure and its rotation are the same shape and size, so they are congruent. The figure may be turned varying degrees in different directions—clockwise or counterclockwise. To keep track of how far an object turns, follow one point on the object as it rotates.

Rotating a figure 90° clockwise

The diagram below shows a triangle being rotated 90° clockwise around point C. You could also look at it as being rotated 270° counterclockwise around point C.

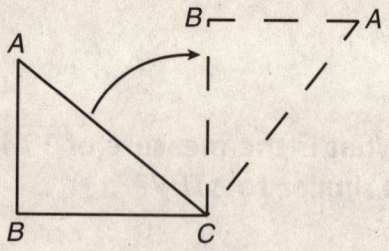

Rotating a figure 270° clockwise

The diagram below shows a triangle being rotated 270° clockwise around point C. You could also look at it as being rotated 90° counterclockwise around point C.

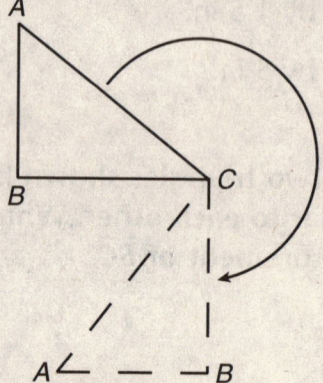

Rotating a figure 180° clockwise

The diagram below shows a triangle being rotated 180° clockwise around point C. This is the same thing as being rotated 180° counterclockwise around point C.

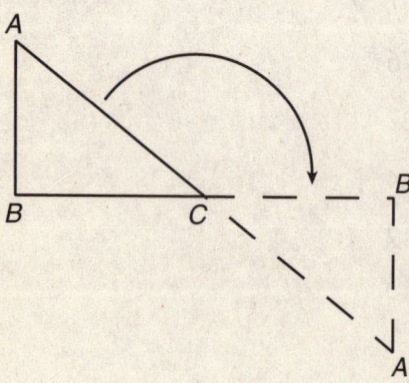

Rotating a figure 360°

When a figure is rotated 360°, it returns to its original position.

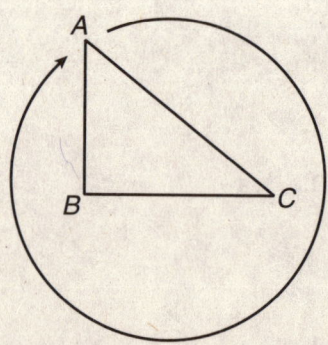

Directions: Read each question carefully. Circle the correct answer choice for each question.

1   Melissa makes a collage out of leftover craft paper. She is trying to decide which way to hang the collage. Which of the following shows Melissa's collage after she turns it 90° clockwise?

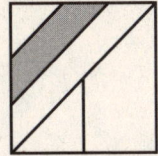

A.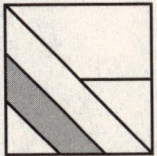

B.

C.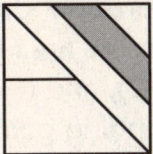

D.

2   Helene is playing a game with a spinner. The pointer ends up pointing 180° counterclockwise from its original position. Which of the following shows the position of the pointer after the spin?

A.

B.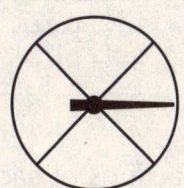

C.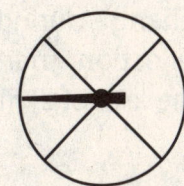

D.

MILE 30: REFLECTIONS AND TRANSLATIONS

Reflections and translations are changes in the positions of geometric figures. Like rotations, they are transformations in which the original figure and the transformed figure are congruent. Let's see how they work.

A **reflection** is a change in position of a geometric figure similar to a mirror image. Figures that are reflected are "flipped" across a line. The reflection of the figure often occurs over the x-axis or y-axis. The line that separates original image and its reflection is called the **line of symmetry** and is exactly in between the images. Observe the three triangles below.

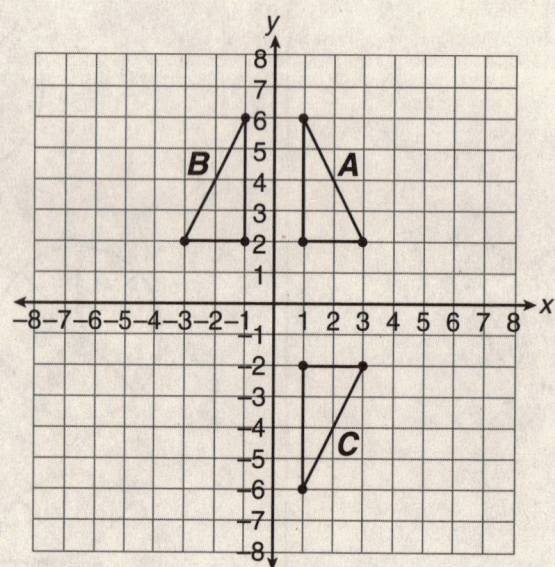

Triangles A and B are reflections of each other over the y-axis. They are mirror images of each other, and the y-axis acts as their line of symmetry. Notice how the corresponding vertices of each triangle change during this reflection. In triangle A, the vertices are (1, 6), (1, 2), and (3, 2). In its reflection, triangle B, the vertices are (−1, 6), (−1, 2), and (−3, 2). When the y-axis acts as the line of symmetry, the x-coordinates of corresponding points change signs.

Triangles A and C are reflections of each other over the x-axis. They are mirror images of each other, and the x-axis acts as their line of symmetry. Again, notice how the corresponding vertices of each triangle change during reflections. In triangle A, the vertices are (1, 6), (1, 2), and (3, 2). In its reflection, triangle C, the vertices are (1, −6), (1, −2), and (3, −2). When the x-axis acts as the line of symmetry, the y-coordinates of corresponding points take the opposite sign.

A **translation** changes the position of a geometric figure by sliding the entire figure to another location. Observe the three figures below.

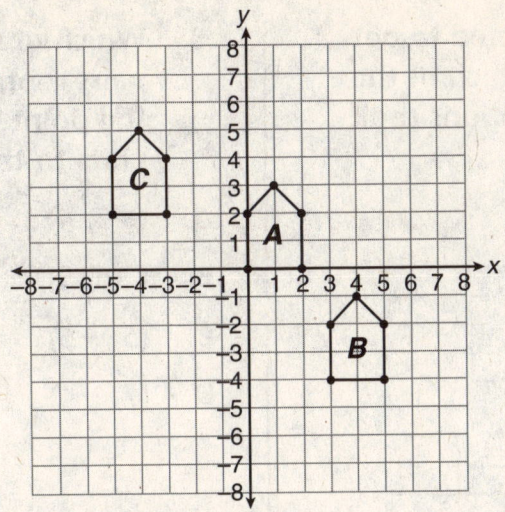

If pentagon *A* is the original shape, then pentagon *B* is a translation of pentagon *A*. Notice how all the corresponding points in *A* move three units to the right (+3) and four units down (–4). The top point of pentagon *A*, (1, 3), becomes (1 + 3, 3 – 4), or (4, –1), in pentagon *B*. Slides to the left or right will change the *x*-coordinate of each point. Slides up or down will change the *y*-coordinate of each point. All the points in a translation will move similarly. Look at other points in pentagons *A* and *B* to make sure.

Pentagon *C* is also a translation of pentagon *A*. Notice how pentagon *C* slides to the left five units (–5) and up two units (+2). In fact, pentagons *A*, *B*, and *C* are all translations of each other because they can slide from one position to another.

You may see more than one transformation occurring to an object. When this is the case, avoid confusion and consider only one transformation at a time. Remember that a translation, unlike a reflection or a rotation, doesn't affect the orientation of the shape. Also, all figures in a translation maintain their shape, but shapes are inverted during the process of reflection.

Directions: In the grid below, first reflect the object across the *y*-axis then translate the object down 6 units.

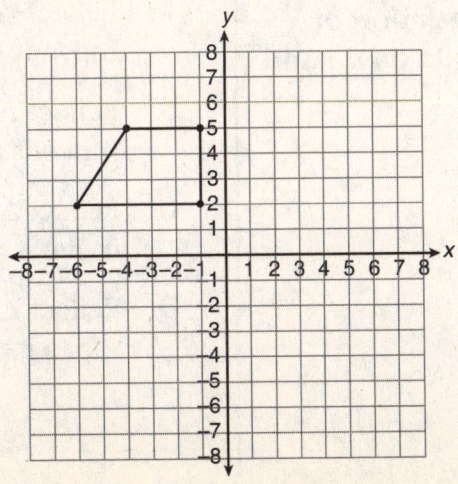

Mile 30: Reflections and Translations

Directions: Read each question carefully. Circle the correct answer choice for each question.

1 △*XYZ* was a transformed to get the image △*MNO*. Which of the following transformations took place?

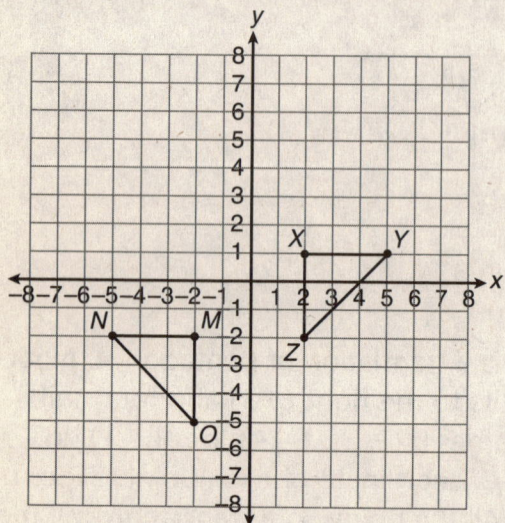

A. Move △XYZ up 3 units and then reflect it over the *x*-2axis.

B. Move △XYZ up 3 units and then reflect it over the *y*-axis.

C. Move △XYZ down 3 units and then reflect it over the *y*-axis.

D. Move △XYZ down 3 units and then reflect it over the *x*-axis.

2 Which of the following coordinate pairs represents a translation of the point (3, −5) 2 units down and 3 units to the right?

A. (0, −7)

B. (5, −2)

C. (5, −8)

D. (6, −7)

3 Which of the following coordinate pairs represents a translation of the point (−2, 0) 5 units up and 4 units to the left?

A. (−7, 4)

B. (−6, 5)

C. (5, −6)

D. (7, −4)

4 △*STU* was transformed to get the image △*PQR*. Which of the following transformations took place?

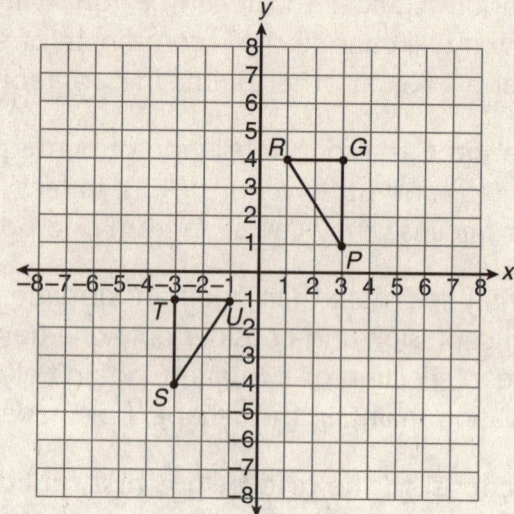

A. Reflect △STU over the *x*-axis and move it up 5 units.

B. Reflect △STU over the *y*-axis and move it up 5 units.

C. Reflect △STU over the *y*-axis and move it down 5 units.

D. Reflect △STU over the *x*-axis and move it down 5 units.

5 △*XYZ* is translated so that the image of point *X* is at (2, 1). What are the coordinates of the image of point *Y*?

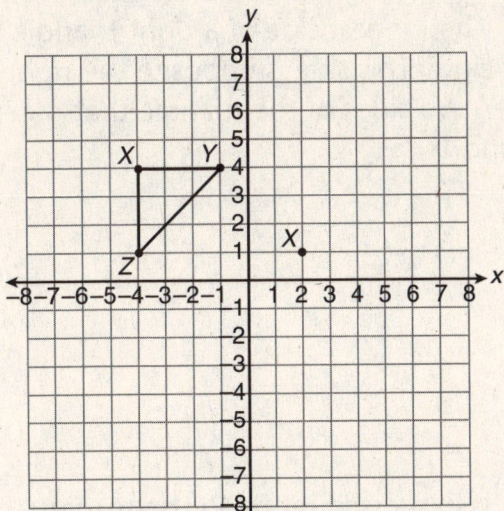

A. (−2, −1)

B. (2, −2)

C. (5, 1)

D. (5, 2)

6 Which of the following coordinate pairs represents a reflection of the point (5, −5) across the *x*-axis?

A. (−5, −5)

B. (0, −5)

C. (5, 0)

D. (5, 5)

7 Quadrilateral *MNOP* is translated so that the image of point *P* is at (−5, 0). What are the coordinates of the image of point *N*?

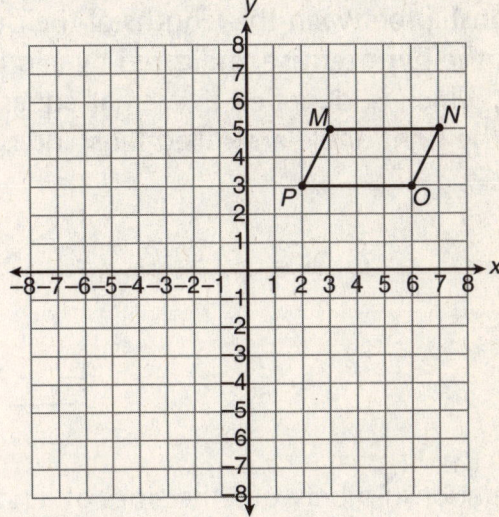

A. (2, 3)

B. (2, 0)

C. (−1, 0)

D. (0, 2)

8 Which of the following coordinate pairs represents a reflection of the point (2, 4) across the *y*-axis?

A. (−2, 4)

B. (0, −4)

C. (2, −4)

D. (4, 2)

MILE 31: PYTHAGOREAN THEOREM

A triangle with a 90° angle is called a right triangle. Right triangles show a unique relationship between the lengths of their three sides. The longest side of a right triangle is called the **hypotenuse,** indicated by c in the figure below. This side is opposite the right angle, which is identified by a small square. The other two sides of the triangle that meet to form the right angle are called **legs,** indicated by a and b.

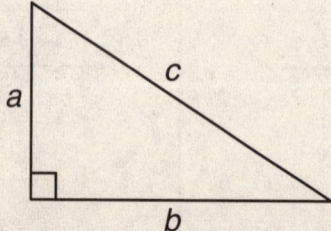

The relationship between the sides of a right triangle is expressed in the **Pythagorean theorem.**

$$a^2 + b^2 = c^2$$

The Pythagorean theorem states that the sum of the squares of the lengths of the legs is equal to the square of the length of the hypotenuse in a right triangle.

Use the Pythagorean theorem to find the length of the hypotenuse of the triangle below.

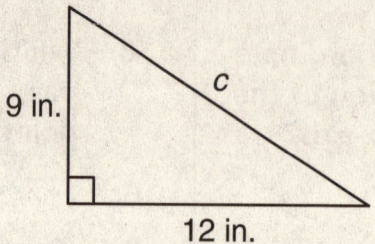

To find the hypotenuse, c, all you have to do is plug in the numbers for sides a and b.

$$9^2 + 12^2 = c^2$$

$$81 + 144 = c^2$$

$$c^2 = 225$$

$$c = \sqrt{225}$$

$$c = 15 \text{ inches}$$

You've just used the Pythagorean theorem to figure out that the length of the hypotenuse is 15 inches! As long as you know the lengths of two of the sides of a right triangle, you can use the Pythagorean theorem to find the length of the third side.

Directions: Figure out how far Katie and Ricardo have to travel to get to various places throughout the day. Solve each problem using the Pythagorean theorem and a calculator. If you get an answer that is not a whole number, round it to the nearest tenth. When you are done, add all of the separate distances to get the total distance they traveled.

1.

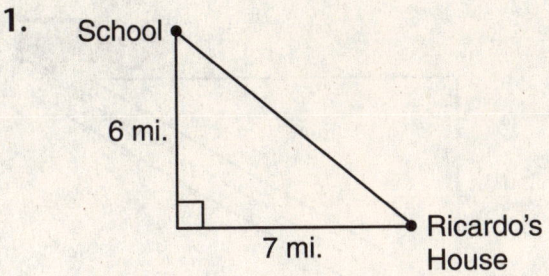

Katie and Ricardo were late for school, so Ricardo's mother drove them directly from Ricardo's house to the school. What is the shortest distance from Ricardo's house to the school? _____

2.

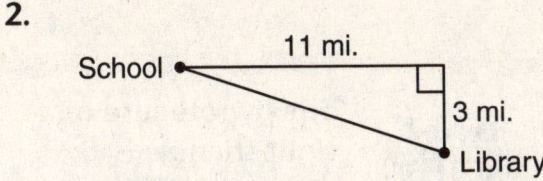

After school, Katie and Ricardo took the bus directly to the public library for a study session. If the bus takes the shortest route from the school to the library, how far did Katie and Ricardo travel?

_____

3.

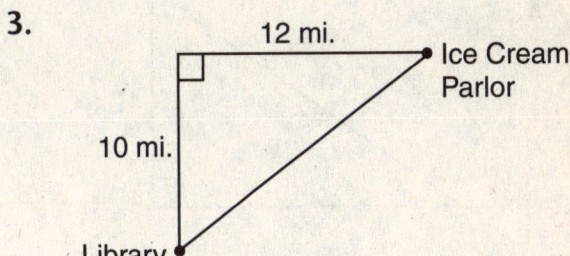

After their study session, Katie and Ricardo took another bus to Armando's Ice Cream Parlor for a snack. The bus traveled in a straight line from the library to the ice cream parlor. How far did they ride? _____

4.

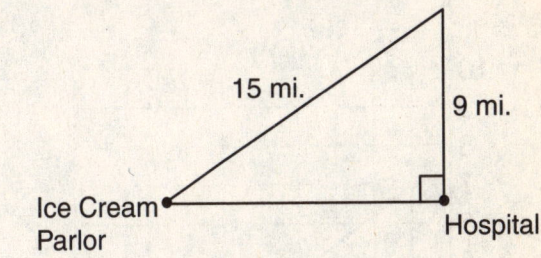

When they were done with their ice cream, Katie's father picked them up and drove them to the local children's hospital where they volunteered one afternoon per week. The diagram above shows the distance they drove from the ice cream parlor to the hospital. How far did they travel?

_____

5. What is the total distance that Katie and Ricardo traveled throughout the whole day?

_____

Directions: Read each question carefully. Circle the correct answer choice for each question.

1 What is the length of the unlabeled leg in the triangle below?

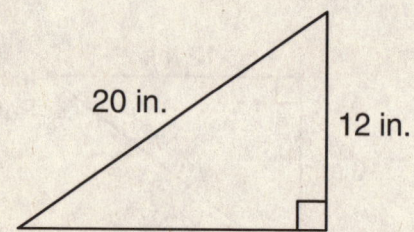

20 in.　12 in.

A.　16 in.

B.　15 in.

C.　12 in.

D.　10 in.

2 What is the length of side y in the triangle below?

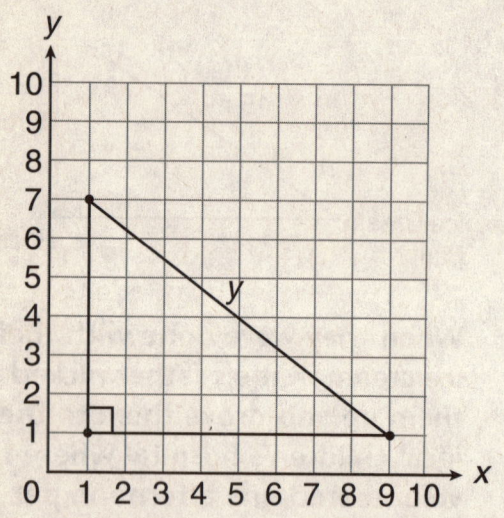

A.　18 units

B.　14 units

C.　12 units

D.　10 units

3 Two cars are shown on a highway below. Which of the following is the distance between the two cars?

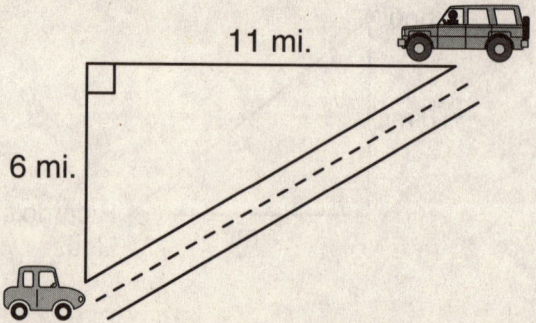

11 mi.

6 mi.

A.　12 mi.

B.　12.5 mi.

C.　13 mi.

D.　13.5 mi.

Tip

The hypotenuse of a right triangle is always *longer* than each of the legs of the triangle. If your calculations result in a hypotenuse that is shorter than a leg of the triangle, you should redo the math using the information given in the question.

4 Which of the following is the measure of the hypotenuse in the triangle below?

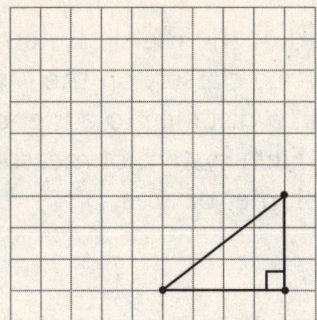

A. 2 units

B. 3 units

C. 5 units

D. 6 units

5 Which of the following expressions is equal to the length of $\overline{JL}$?

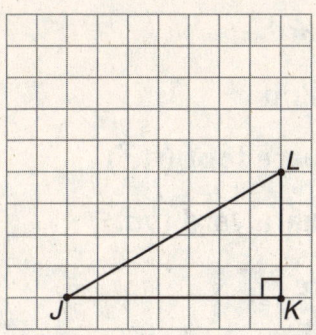

A. $7 + 4$

B. $7^2 + 4^2$

C. $\sqrt{(7 + 4)}$

D. $\sqrt{(7^2 + 4^2)}$

6 Which of the following is equal to the length of $\overline{LJ}$?

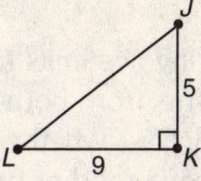

A. $\sqrt{45}$

B. $\sqrt{106}$

C. 45

D. 106

7 Which of the following is the measure of the unlabeled leg in the triangle below?

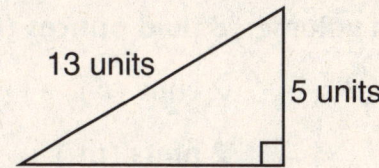

A. 8 units **B.** 10 units

C. 12 units **D.** 18 units

8 Which of the following is the correct expression for the length of the unlabeled leg in the triangle below?

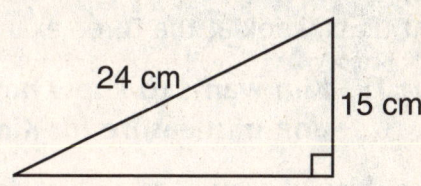

A. $24 - 15$

B. $24^2 - 15^2$

C. $\sqrt{(24^2 - 15^2)}$

D. $\sqrt{(24^2 + 15^2)}$

MILE 32: MEASUREMENT

Measurements involve units from either the American system or the metric system. You will see measurements from both systems in your math and science classes, on the MCAS tests, and in everyday life—whether it's on a speedometer in a car, a distance to a friend's house, or the amount of a food or beverage you buy from a store. Each system of measurement has its own units to calculate length, weight/mass, liquid volume, or area.

American system

Length:	12 inches (in.)	=	1 foot (ft.)
	3 feet	=	1 yard (yd.)
	5,280 feet	=	1 mile (mi.)
Weight/Mass:	16 ounces (oz.)	=	1 pound (lb.)
	2,000 pounds	=	1 ton
Liquid volume:	8 fluid ounces (fl. oz.)	=	1 cup
	2 cups (c.)	=	1 pint
	2 pints (pt.)	=	1 quart
	4 quarts (qt.)	=	1 gallon
Area:	144 square inches (in.^{2})	=	1 square foot (ft.^{2})
	9 square feet	=	1 square yard (yd.^{2})
	43,560 square feet	=	1 acre
	640 acres	=	1 square mile (mi.^{2})

You may be asked to find the best unit of measure for different objects or convert between different units. Look at the three examples below.

Example 1: **Kim wants to know how heavy her dog is. What is the most appropriate unit of measure for Kim to use?**

To figure this out, you need to first decide on the kind of measurement. Kim wants to know the weight of her dog and weight is measured in ounces, pounds, or tons. If ounces are used, the dog will be many, many ounces. If tons are used, the dog will be a very small fraction of a ton. So the most appropriate unit is the pound.

Example 2: **What is the distance of a mile in yards?**

Look at the chart above. There is no direct conversion between miles and yards, but 5,280 feet = 1 mile, and 3 feet = 1 yard. To figure out how many yards there are, simply divide 5,280 by 3 to get 1,760 yards.

Example 3: **How many fluid ounces are there in a gallon?**

If you look at the chart, you won't find a direct conversion from ounces to gallons, but there is enough information there to help you. There are 4 quarts in a gallon and 2 pints in each quart, so there are $4 \times 2 = 8$ pints in a gallon. There are 2 cups in each pint, so there are $8 \times 2 = 16$ cups in a gallon. Finally, there are 8 fluid ounces in each cup, so there are $16 \times 8 = 128$ fluid ounces in a gallon.

Metric system

Length:	**The meter is the basic unit of length.**		
	10 millimeters (mm)	**=**	**1 centimeter (cm)**
	100 centimeters	**=**	**1 meter (m)**
	1000 meters	**=**	**1 kilometer (km)**
Weight/Mass:	**Grams are the basic unit of mass.**		
	1000 milligrams (mg)	**=**	**1 gram (g)**
	1000 grams	**=**	**1 kilogram (kg)**
Liquid volume:	**Liter is the basic unit of liquid volume.**		
	1000 milliliters (mL)	**=**	**1 liter (L)**
Area:	**10 000 square centimeters (cm^2)**	**=**	**1 square meter (m^2)**
	1 000 000 square meters (m^2)	**=**	**1 square kilometer (km^2)**

The units of the **metric system** are related by the powers of 10. Conversion within this system is easier to perform because you only have to multiply or divide by a power of 10. Look at the example below.

Example 4: **Samuel wants to buy a replacement part for his computer. He knows that the part measures 0.08 meters. The catalog he is ordering from only lists the sizes in millimeters. What size would Samuel order?**

Look at the chart above. There is no direct conversion from meters to millimeters, but you know that there are 100 centimeters in a meter, so Samuel's part measures $0.08 \times 100 = 8$ centimeters. There are 10 millimeters in 8 centimeters, so he would order a $10 \times 8 = 80$ millimeter part.

Example 5: **What is 170 grams in kilograms?**

There are 1000 grams in one kilogram, so you know that 170 grams will be a lot less than 1 kilogram. Divide 170 by 1,000 to get 0.17 kilograms.

Directions: Read each question carefully. Circle the correct answer choice for each question.

1   A meter is the most appropriate unit of measure for which of the following?

A.   the length of a piece of fabric

B.   the area of a piece of fabric

C.   the amount of liquid in a bottle

D.   the volume of liquid in a bottle

2   Sara used square inches as the unit for a measurement that she made. Which of the following did Sara measure?

A.   the length of her room

B.   the volume of air in a room

C.   the area of a floor tile

D.   the perimeter of a floor tile

3   Which of the following units of measurement is most appropriate to measure the amount of water in a swimming pool?

A.   yards

B.   miles

C.   square feet

D.   liters

4   A cubic meter can be used to measure which of the following?

A.   area

B.   volume

C.   height

D.   weight

5   Jamila bought 17 pints of juice to make punch for a party. How many gallon jugs can she completely fill with punch?

A.   2

B.   4

C.   6

D.   8

Many metric units are given names whose prefixes explain the value of the unit. For example, "kilo-" means one thousand, "centi-" means one-hundredth, and "milli-" means one-thousandth.

6 A local campground covers about 3,000 acres of land. About how many square miles is the campground?

A. 2

B. 5

C. 10

D. 30

7 Kevin drew a map of his neighborhood that covered 3 square miles. About how many acres did his map cover?

A. 10

B. 1,000

C. 2,000

D. 5,000

8 An airplane is traveling at 30,000 feet. About how many miles in the air is the airplane?

A. 30

B. 15

C. 10

D. 6

9 Which of the following units of measurement is most appropriate to measure the amount of cardboard used to make a large storage box?

A. cubic inches

B. pints

C. square feet

D. yards

10 The public library is 1200 meters from the school. How many kilometers is this?

A. 1.2

B. 12

C. 100

D. 1000

When converting from one system of measurement to another, it helps to learn approximations or "ballpark" figures. Here are some common approximations.

1 gram is about the weight of a common paper clip

1 foot is about 30 centimeters.

1 ton is about 1000 kilograms.

1 mile is about 1.5 kilometers.

1 kilogram is about 2 pounds.

1 meter is about a few inches more than 1 yard.

1 liter is about 1 quart.

1 inch is about 2.5 centimeters.

Mile 33: Perimeter

Perimeter is the measure of the border around a two-dimensional shape or polygon, such as triangle, a square, or a rectangle. The perimeter is the sum of all the sides of a polygon.

For an irregular shape, the lengths of each side will most likely be given to you or the diagram will provide enough information for you to find the lengths of each side. Then simply add up the lengths of each side to find the perimeter.

Example 1:   **Find the perimeter of the shape below.**

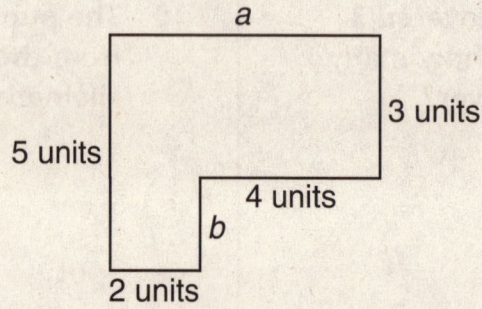

You are only given some of the side lengths for the figure above. To find the perimeter, first find the lengths of the missing sides. This shape has only right angles, so you know that length *a* has to be equal to the sum of the opposite lengths 2 and 4. Length *a* is equal to 6 units. You can also see that length *b* has to be equal to the difference between 5 and 3. Length *b* is equal to 2 units. Now find the perimeter by adding up the lengths of all the sides: 6 + 3 + 4 + 2 + 2 + 5 = 22 units.

You can find the perimeter of regular polygons if you only know the length of one side. Remember that regular polygons have congruent sides, so if you know the length of one side you know the length of all the sides.

Example 2:   **One side of an equilateral triangle measures 15 inches. What is the perimeter of the triangle?**

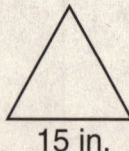

15 in.

Although only one side of the triangle above is given, you know that an equilateral triangle has three congruent sides. So the lengths of the other sides also measure 15 inches each. Add the lengths of all the sides to get the perimeter: 15 + 15 + 15 = 45 inches. Because you are adding the same number multiple times, you can also multiply the number of sides by the length of one side: 3 × 15 = 45 inches.

Directions: Read each question carefully. Circle the letter next to your answer choice.

1   What is the perimeter of the park shown below?

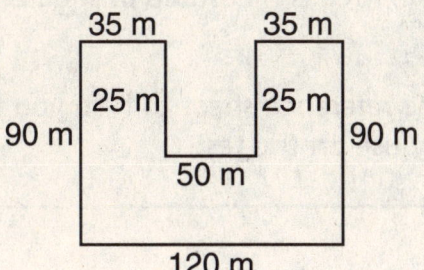

A. 420 meters

B. 445 meters

C. 470 meters

D. 495 meters

2   If a square measures 19 inches on one side, what is its perimeter?

A. 76 inches

B. 80 inches

C. 190 inches

D. 361 inches

3   If a rectangle is 79 inches long and 35 inches wide, what is its perimeter?

A. 114 inches

B. 228 inches

C. 1,225 inches

D. 2,765 inches

4   A regular hexagon measures 60 cm on one side. What is the perimeter of the hexagon?

A. 60 cm

B. 120 cm

C. 360 cm

D. 600 cm

5   Find the perimeter of the shape shown below.

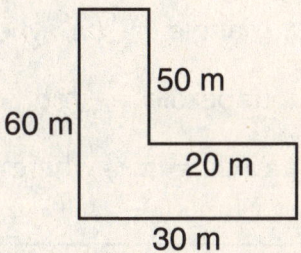

A. 150 m

B. 160 m

C. 170 m

D. 180 m

6   The perimeter of a square playground equals 260 yards. What is the length of one side of the playground?

A. 60 yards

B. 65 yards

C. 200 yards

D. 1,040 yards

The amount of surface covered by a two-dimensional figure is its **area.** It may be a plane figure, like a polygon, or a curved figure, such as a circle. Area is measured in square units, often represented as cm^2 or $in.^2$.

The Grade 8 MCAS Math test lists some formulas on the reference sheet to help you find area. Here are some of the important formulas you may use on the test.

Area of a square $= s^2$, where s is the length of one of its sides

Area of a rectangle $= bh$, where b is its base and h is its height, or $l \times w$, where l is its length and w is its width

Area of a triangle $= \frac{1}{2}bh$, where b is its base and h is its height

Area of a trapezoid $= \frac{1}{2}h(b_1 + b_2)$, where b_1 and b_2 are the bases and h is its height

Area of a circle $= \pi r^2$, where r is its radius. Remember that π is about 3.14.

Example: **The floor in a school's gymnasium is in the shape of a rectangle and measures 81 ft. by 42 ft. What is the area of the floor in square yards?**

You are asked to find the area in square yards, but the length and width are given in feet. Before you multiply the length by the width, divide by 3 to convert the given measurements into yards: 81 ft. ÷ 3 = 27 yd., and 42 ft. ÷ 3 = 14 yd.

Now that you have the dimensions of the gym in yards, you can go ahead and find the area by multiplying the length by the width. 27 yd. × 14 yd. gives you an area of 378 $yd.^2$.

Directions: Read each question carefully. Write your answer in the space below the question or circle the letter next to your answer choice.

1 Dawn is painting a mural on a piece of canvas for an art project. She bought a piece of canvas that is 48 inches by 72 inches. The canvas cost $3.50 per square foot. How much did Dawn pay for the canvas?

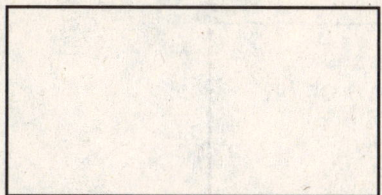

2 Tiles cost $1.50 per square foot. How much will it cost to tile a square countertop that is 2 yards long and 2 yards wide?

A. $6

B. $9

C. $36

D. $54

3 The diameter of a circle is 8 feet. What is the area of the circle?

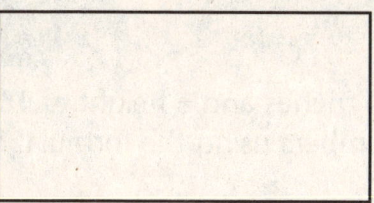

4 The perimeter of the figure below is 42 units. What is the area of one of the small squares?

A. 9 square units

B. 6 square units

C. 3 square units

D. 1 square unit

5 The perimeter of the figure below is 52 units. What is the area of one of the small squares?

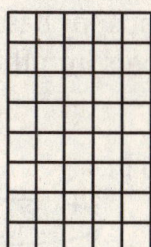

A. 8 square units

B. 5 square units

C. 4 square units

D. 2 square units

Volume is the amount of space that an object occupies. It is a common measurement that can be useful in many everyday situations. For example, say you want to compare bottles of juice at a grocery store. One way to compare them would be to look at the volume of the bottles. The volume is a measure of the space inside the bottle or how much a bottle can hold. Looking at the volume of the bottles can tell you which bottle holds more juice.

Here are the volume formulas you'll need to know as you travel through this mile.

Rectangular prism:	$V = lwh$ or Bh
Cylinder	$V = \pi r^2 h$
Cone	$V = \frac{1}{3}\pi r^2 h$
Cube	$V = s^3$

Remember that l stands for length, w stands for width, h stands for height, r stands for radius, and B stands for base area. Use 3.14 for the value of π. All you have to do is plug the correct numbers into these formulas to answer questions about volume.

Example:

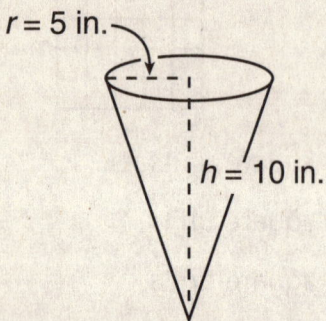

$r = 5$ in.

$h = 10$ in.

The figure above tells you that the cone has a radius of 5 inches and a height of 10 inches. To figure out the volume of the cone, just plug in the numbers using the formula!

$V = \frac{1}{3}\pi(5^2)(10)$

$V = \frac{1}{3}\pi(25)(10)$

$V = \frac{1}{3}\pi(250)$

$V = 83\frac{1}{3}\pi$

If $\pi = 3.14$, the volume of the cone is approximately $261\frac{1}{3}$ cubic inches. The term cubic inches can also be expressed as "in.^{3}."

Directions: Bradley and Devon are watching their sister's soccer game one summer afternoon. Bradley has a box of juice and an ice-cream cone, as pictured below. Figure out the volume of the ice cream cone and box of juice. (Use 3.14 for π.)

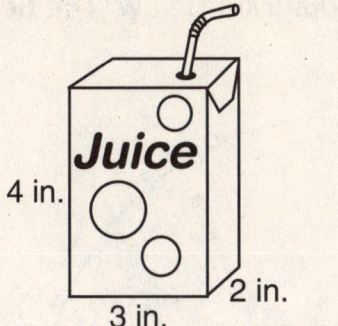

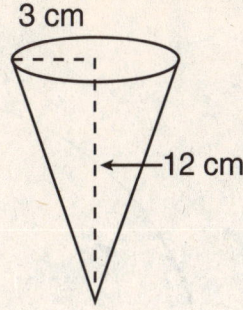

1 Bradley starts out with a full box of juice. What is the volume of the box?
_____

2. What is the volume of Bradley's ice-cream cone? _____

3. Devon has a juice box too. If the height of Devon's juice box is one-quarter less than Bradley's, what is the volume of the Devon's juice box? _____

4. How does the volume of Devon's juice box compare to the volume of Bradley's juice box? _____

5. Devon also has an ice-cream cone. He has a waffle cone that has a height that is four-thirds as big as the cone Bradley has. If the radius of the waffle cone is the same as Bradley's cone, what is the volume of Devon's ice-cream cone?
_____

6. Compare the volumes of the two cones. _____

Directions: Read each question carefully. Write your answer on a separate sheet of paper or circle the correct answer choice for each question.

1   The bottom of a can is shown in the diagram below. The height of the can is 10 centimeters.

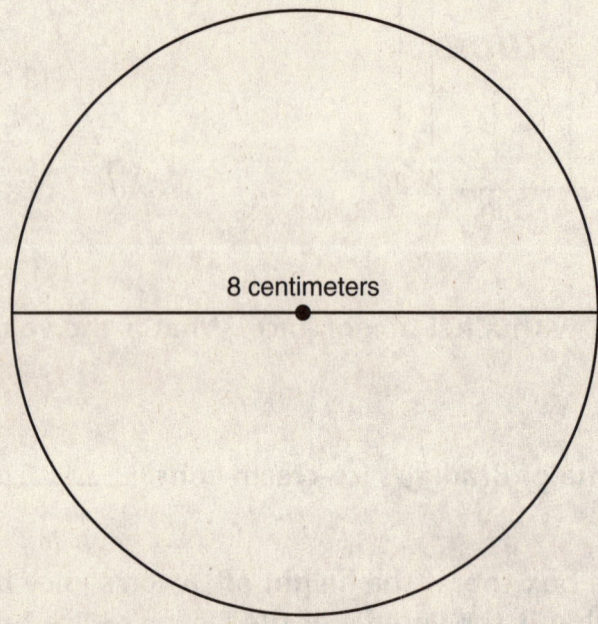

8 centimeters

a. What is the radius of the can in centimeters?

b. What is the volume of the can? Show your calculations.

c. What is the volume of a can that is half the diameter and half the height of the can in the diagram? Show your calculations.

d. What is the ratio of the volume of the larger can in part *b* to the volume of the smaller can in part *c*?

2 What is the volume of the prism in the diagram below?

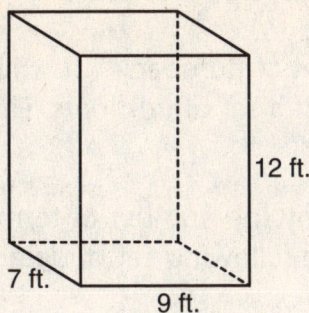

12 ft.

7 ft.

9 ft.

A. 84 ft.³

B. 108 ft.³

C. 756 ft.³

D. 2,268 ft.³

3 What is the volume of a cone that has a radius of 30 inches and a height of 15 inches?

A. 14,130 in.³

B. 1,413 in.³

C. 450 in.³

D. 42,390 in.³

4 What is the volume of the cube in the diagram below?

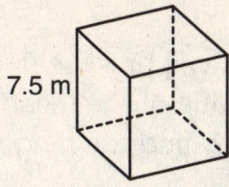

7.5 m

A. 22.5 m³

B. 56.25 m³

C. 176.625 m³

D. 421.875 m³

5 What is the volume (to the nearest tenth) of a cone that has a radius of 4 ft. and a height of 3.5 ft.?

A. 12.3 ft.³

B. 49.0 ft.³

C. 58.6 ft.³

D. 175.8 ft.³

MILE 36: MEAN, MEDIAN, MODE, AND RANGE

In this mile you will practice finding the mean of a given set of numbers. You will also compare the value of the mean to the median and mode of a set of numbers. Finally, you will learn how to find the range of a number set.

The **mean** is the sum of the terms in a set of data divided by the number of terms in the set. Another word for the mean is the **average**. Look at the following set of data: {6, 21, 23, 26, 25, 5, 24, 30}.

There are eight numbers in the set above. To find the mean you would do the following calculations:

$$\frac{(6 + 21 + 23 + 26 + 25 + 5 + 24 + 30)}{8} = \frac{160}{8} = 20$$

The mean of this given set is 20.

The **median** is the middle number in a set when the numbers are organized in numerical order. You don't have to do any calculations to find the median. Just put the numbers in order and find the one in the middle. Consider the following set of data: {19, 34, 2, 41, 20, 36, 8}.

The numbers above are out of order, so before you find the median you have to put them in order: 2, 8, 19, 20, 34, 36, and 41. Because 20 is the middle number, it is the median of this set.

When there is an even number of numbers in a set, there will be no clear middle number. Look at the set {10, 12, 14, 20}. To find the median of this set of numbers, find the two middle numbers, 12 and 14, and take the mean of those numbers. (12 + 14) ÷ 2 = 13, so the median is 13.

The **mode** is the number that appears most often in a set of numbers. You also don't have to do any calculations to find the mode. Just identify the number that appears most often in the set. Look at the set {45, 236, 675, 53, 2, 45, 236, 54, 45}.

The mode in the above set of numbers is 45, which appears three times.

The **range** is the difference between the greatest and least values in a set of numbers. To find the range, subtract the least number in the set from the greatest number. The difference is the range. In the set of numbers {5, 8, 11, 17, 20, 25}, the least number is 5, and the greatest number is 25. The difference is (25 − 5), or 20. The range is 20.

Directions: Solve the problems at the bottom of the page and then write the answers into the corresponding spaces in the cross-number puzzle below.

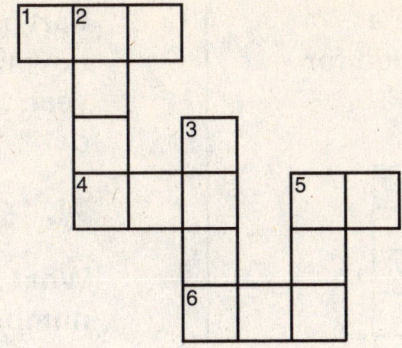

Across

1. What is the sum of the mode and mean of the following set of numbers: {35, 64, 40, 65, 67, 64, 50, 55}?

4. What is the median of the set {4, 10, 222, 503, 510, 548, 643}?

5. What is the range of the set {1, 15, 17, 3, 11, 8}?

6. What is the mode of the set {621, 611, 621, 631, 611, 631, 611}?

Down

2. What is the mode of the set {2,048; 7,645; 1,045; 2,045; 3,485; 1,045; 7,644}?

3. What is the sum of the mean and median of the following set of numbers: {648, 664, 656, 683, 709}?

5. What is the mean of the following set of numbers: {118, 162, 179, 194, 202}?

Directions: Read each question carefully. Circle the letter next to your answer choice for each question.

1 The temperatures inside a greenhouse were recorded for seven days.

Day	Temperature
1	88°F
2	90°F
3	89°F
4	87°F
5	97°F
6	95°F
7	91°F

What was the mean temperature during the week?

A. 89°F

B. 90°F

C. 91°F

D. 96°F

2 Cassandra's science quiz scores for the semester so far are 85, 89, 21, 89, 88, and 90. Which of the following is the mean of her scores?

A. 60

B. 77

C. 89

D. 95

3 Karen asked 7 different students how many movies they saw last year. She recorded the following data:

11, 14, 6, 4, 9, 22, 18

What is the median of the set of numbers above?

A. 11

B. 9

C. 3

D. 1

4 Jill made 12 phone calls to her friends over the last week. Below is a list of the length of each call in minutes.

24, 8, 15, 28, 5, 12, 33, 41, 30, 8, 11, 35

What is the range, in minutes, of the lengths of Jill's phone calls?

A. 31

B. 36

C. 41

D. 46

 You don't need to perform any mathematical operations when you are asked to find the mode. To find the mode, just choose the number that appears most often.

5. A restaurant served a different number of people for each hour it was open. The restaurant was open for 10 hours and the list below shows the number of people it served each hour.

12, 44, 51, 60, 55, 39, 41, 32, 20, 9

What is the range of customers it served hourly?

A. 9

B. 39

C. 51

D. 55

6. Huang made a list of the shoe sizes of some of his classmates as part of a math project. The following are Huang's data:

8, 7.5, 8.5, 5, 6.5, 7, 8.5, 8, 9, 11, 10.5, 9, 6.5, 6.5, 7, 8.5, 4, 8.5

What is the mode of Huang's data?

A. 6.5

B. 7

C. 8

D. 8.5

7. A fishery releases small fish into a lake each year. The following is a sample of the mass of the fish in grams:

22, 27, 28, 30, 26, 24, 23, 28

According to the sample, what which of the following has the greatest value?

A. mode

B. median

C. mean

D. range

8. Five friends went shopping and spent the following amounts:

$38, $15, $55, $12, $15

What is the range of the amounts that they spent?

A. $15

B. $27

C. $43

D. $55

9. Trent's practice times for a ski run are shown below in seconds.

121   116   125   116   120   119

What is the median of Trent's practice times?

A. 119 seconds

B. 119.5 seconds

C. 120 seconds

D. 120.5 seconds

MILE 37: COMBINATIONS

Figuring out the number of possible combinations of two or more things is like mixing and matching. Pretend that Uncle Bernie is deciding what to wear. He has two shirts and three ties to mix and match.

How many possible combinations of *one shirt* and *one tie* can Uncle Bernie choose from? Each shirt can go with one of three different ties, so there are three possible combinations for each shirt. Because there are two shirts, there are six possible combinations: $3 \times 2 = 6$.

Another way to find all the possible combinations is to make an organized list or a tree diagram. The organized list below shows all six possible combinations.

white shirt/black tie	striped shirt/black tie
white shirt/gray tie	striped shirt/gray tie
white shirt/polka-dot tie	striped shirt/polka-dot tie

Directions: Read the short passage below and figure out how many different types of pizza can be ordered from Sam's Pizzeria.

Sam's Pizzeria offers a choice of three toppings: sausage, peppers, and extra cheese. It is possible to order a pizza with no toppings, one topping, two toppings, or all three toppings. How many different types of pie you can order from Sam's? Write all of the different possible combinations below.

How many different pies is it possible to choose from? _____

Directions: Read each question carefully. Write your answer in the space provided or circle the correct answer choice for each question.

1   Shawna has the two sets of cards shown below. She selects one card from each set and then multiplies the numbers.

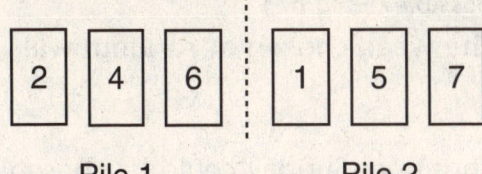

Pile 1   Pile 2

Which of the following is the set of all possible outcomes?

A.   1, 2, 4, 5, 6, 7

B.   2, 4, 6, 10, 14

C.   2, 4, 6, 10, 14, 20, 28, 30, 42

D.   2, 4, 6, 8, 10, 12, 14, 16, 18, 20

2   Judy has a number cube with sides labeled 1 through 6. She rolls a 3 on the cube. If she rolls the cube two more times and gets two even numbers, show a list of the possible outcomes for all three rolls.

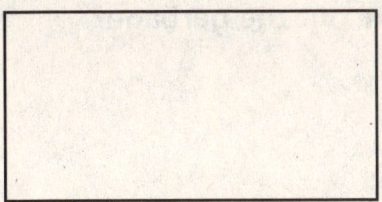

3   Kyle has the spinner in the diagram below. If he spins the spinner twice, what is the probability that Kyle will spin a 2 followed by a 3?

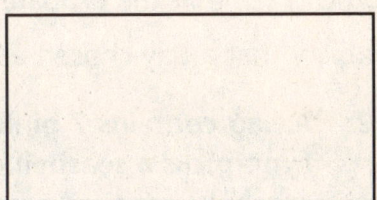

4   Duane, Tasha, and Sydney are in a cross-country race. There are no ties in this race. In how many ways can they finish?

A.   2

B.   3

C.   4

D.   6

MILE 38: PROBABILITY

Probability is the chance that a certain event will happen. It is defined as the ratio of the number of ways a specific event—called a positive outcome—can happen to the total number of possible outcomes. This is the rule for the probability of one event.

$$\text{Probability} = \frac{\text{Number of positive outcomes}}{\text{Total number of possible outcomes}}$$

Example 1:   **What is the probability that a day of the week chosen at random will begin with the letter *S*?**

There are two positive outcomes: Saturday could be chosen or Sunday could be chosen.

There are seven days in a week, so there are seven total possible outcomes. Using the rule above, you can see that the probability is $\frac{\text{Number of days beginning with the letter } S}{\text{Total number of days of the week}} = \frac{2}{7}$.

The probability that a day chosen at random will begin with the letter S is $\frac{2}{7}$.

Example 2:   **A bag contains 7 pink seashells, 6 yellow seashells, and 5 gray seashells. Lynn picks a seashell out of the bag without looking inside. If all the seashells are the same size and shape, what is the probability that she will pick a gray seashell?**

There are 5 gray seashells, so the number of positive outcomes is 5. The total number of seashells in the bag is 5 + 6 + 7, or 18, so the total number of outcomes is 18. Again, using the rule above, you should see that the probability of picking a gray seashell is $\frac{5}{18}$.

Sometimes you will have to convert the ratio into a decimal or percent. To do this, convert the ratio as you would any other fraction.

Example 3:   **Chad uses the spinner below to determine which chore to perform each day. What percent of the time will he take out the garbage?**

There are a total of 4 possible outcomes. The probability that Chad will take out the garbage is 1 out of the 4 chores, or $\frac{1}{4}$. To write this as a percent, divide 1 by 4 to get 0.25 and rewrite to get 25%. Chad will take out the garbage 25% of the time.

Directions: Read each of the possible events listed below and determine its probability based on the given information and your knowledge of the events. State the probability of each outcome in the blank provided as a fraction and as a decimal. (Repeating decimals and irrational numbers can be expressed as approximate values.)

1. An integer from 0 to 9 picked at random is evenly divisible by 3. _____

2. A date picked at random (not from a leap year) is July 14. _____

3. A student randomly selected from a class composed of 10 girls and 10 boys will be a girl. _____

4. A month randomly chosen will begin with the letter *J*. _____

5. A day randomly selected in September will include an 8 in its number. _____

6. A state in the United States that is picked at random will begin with the letter *A*. _____

7. An integer between 1 and 25 picked at random will have a 3 in it. _____

8. A person chosen at random will have a birthday in the month of February (not during a leap year). _____

9. A student randomly selected from a class composed of 18 boys and 16 girls will be a boy. _____

10. A red marble will be picked at random from a bag containing 10 blue marbles, 14 red marbles, 16 green marbles, and 20 black marbles. _____

Directions: Read each question carefully. Write your answer in the space below the question or circle the letter next to your answer choice.

1   Isaiah has a bag with black, green, and blue marbles. If he randomly takes out a marble, the probability of choosing a black marble is $\frac{1}{4}$. The probability of choosing a green marble is $\frac{5}{12}$, and the probability of choosing a blue marble is $\frac{1}{3}$. What is the least possible number of marbles in the bag?

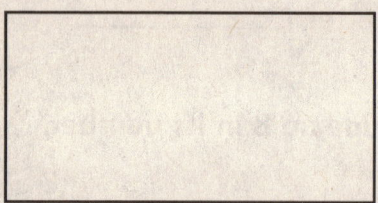

2   Mya bought a pack of erasers. There are green, yellow, and blue erasers in the pack. If she removes an eraser without looking, the probability of choosing a green one is $\frac{1}{5}$. The probability of choosing a yellow one is $\frac{1}{2}$, and the probability of choosing a blue one is $\frac{3}{10}$. If there are 2 green erasers, how many erasers are in the pack?

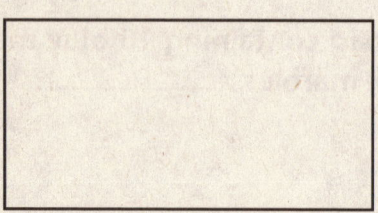

3   When Gina and Vincent cleaned out their mother's car, they found lots of loose change. When they finished, they put the loose change in a jar. In total they found 3 quarters, 24 pennies, and 5 nickels. What is the probability that a coin selected at random from the jar will be a penny?

A. $\frac{4}{5}$

B. $\frac{3}{32}$

C. $\frac{3}{4}$

D. $\frac{24}{40}$

4   Maristella is playing a counting game with her younger brother. She has 13 white beans, 23 red beans, and 25 black beans in a bowl. If her younger brother takes a bean from the bowl without looking, what is the probability that he will get a red bean?

A. $\frac{25}{61}$

B. $\frac{23}{61}$

C. $\frac{23}{48}$

D. $\frac{13}{46}$

5 The number of students that ride the school bus to school in a local district is shown in the table below.

Grade Levels	Number
3–5	225
6–8	128
9–11	212

If one bus-riding student is selected at random, what is the probability that the student will be in a grade higher than the fifth grade?

A. 23%

B. 40%

C. 60%

D. 72%

6 Randall is competing at a gymnastics meet. He has to perform routines on the rings, the parallel bars, and on the floor. His favorite event is the rings. If the order of is routines are randomly determined, what is the probability that his favorite event will be his first event?

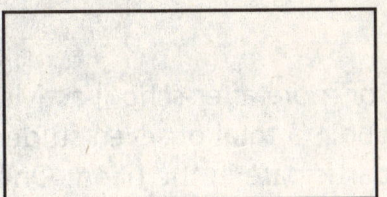

Directions: Write your answer for question 7 on a separate sheet of paper.

7 Sasha and Jerry are playing a game with a set of cards. The face of each card has a color and a number. The colors are orange, red, and blue. There are 3 cards for each color and they are numbered from 1 to 3.

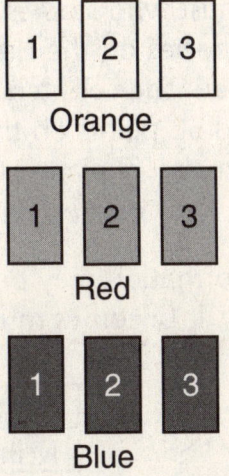

Two cards are selected at random. If one or both of the cards are blue, then the player receives no points. If both cards are either orange or red, then the score is the product of the two cards.

a. Make a list of the possible scores.

b. If Jerry selects an orange card with a 3, what is the probability that the next card he draws will help him score points? Explain your answer.

MILE 39: DATA WITH GRAPHS AND TABLES

The ability to organize data is an important problem-solving skill. Each day you are presented with information that you must sort through before you can draw conclusions. Sorting through all this information is easier if you learn some organizational strategies.

There are a several ways to organize data. One way is to present the data in a **table**. A table is a chart that uses headings, columns, and rows to help organize data. Read the following example:

Olga is a scientist who studies endangered species. For her research, Olga counted endangered species on the Hawaiian Islands. She gathered the following results: On the island of Ni'ihau, she counted 1 bird, 1 mammal, and 6 plants. She counted 13 birds, 2 mammals, and 68 plants on the island of Kaua'i. On the island of Oahu, she counted 7 birds, 2 mammals, 79 plants. On the island of Maui, she counted 12 birds, 2 mammals, 64 plants. Finally, she counted 14 birds, 2 mammals, and 51 plants on the big island of Hawaii.

The above information may be difficult to grasp. But if the information is in a table such as the one below, it becomes much easier to understand.

Island	Birds	Mammals	Plants
Ni'ihau	1	1	6
Kaua'i	13	2	68
Oahu	7	2	79
Maui	12	2	64
Hawaii	14	2	51

Another way to sort data is with a **Venn diagram.** A Venn diagram shows the relationship between sets drawn as circles, where any overlap of circles represents a common element between the sets. For example:

Some students in Ms. Bernosky's class participate in one or more after-school activities. There are a total of five students who participate in the choir, a total of seven students who participate in sports, and a total of seven students who participate in the band. One student participates in all three activities, one student participates in sports and in the band, and one student participates in the choir and in sports.

The information above can be organized as a Venn diagram. The Venn diagram must have three circles: one for choir, one for sports, and one for band. The portions of the circle that overlap show students that participate in more than one activity and therefore belong to more than one group. The Venn diagram on the next page shows the after-school activities of the students in Ms. Bernosky's class.

Students in After-School Activities

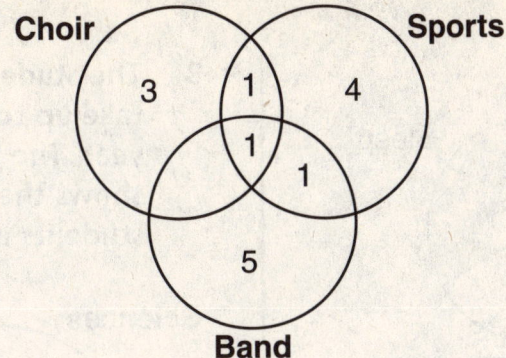

All the numbers in each activity circle add up to the total sums stated before. You can easily see that three students participate in one or more activities. In the regions with no overlap, you can see the number of students that participate in only that activity.

Another way of representing data is to use a **circle graph** (sometimes referred to as a **pie graph**). A circle graph shows comparisons of parts of a whole.

Jill did a survey to find what students did in their spare time and got the following results: 10 liked to read, 4 liked to swim, and 6 liked to do computer related activities. To make a circle graph, Jill would have to determine what part of the circle graph should be used for each activity. First, find the total number of students surveyed: 10 + 4 + 6 = 20 students. Then find the fraction that represents each activity: $\frac{10}{20}$ of the students like to read, $\frac{4}{20}$ of the students like to swim, and $\frac{6}{20}$ of the students like to do computer-related activities.

Next, calculate the number of degrees for each sector of the graph. Because there are 360° in a circle, you can write a proportion to find the number of degrees in each sector.

Reading: $\frac{10}{20} = \frac{180}{360}$ or 180°

Swimming: $\frac{4}{20} = \frac{72}{360}$ or 72°

Computer: $\frac{6}{20} = \frac{108}{360}$ or 108°

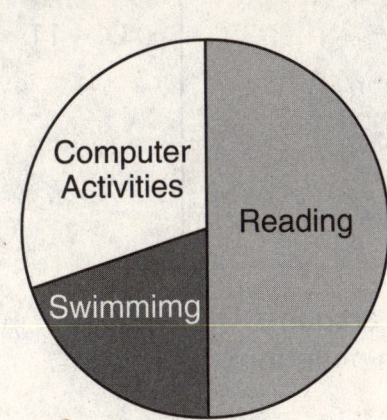

Directions: Read each question carefully. Write your answers on a separate sheet of paper or circle the letter next to your answer choice.

1

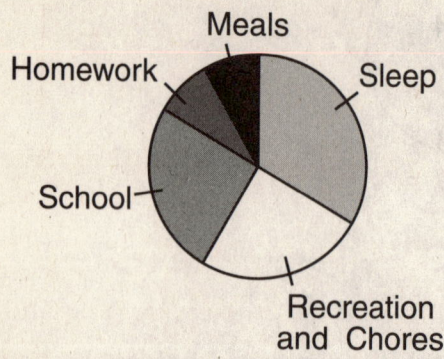

The circle graph above shows the number of hours Margo spends for each activity during one day. Which activity takes up about 8 hours of her day?

A. homework

B. recreation and chores

C. school

D. sleep

2 **Sources of Ocean Pollution**

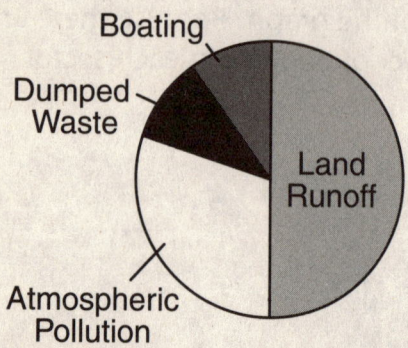

Which of the following accounts for about 30% of ocean pollution?

A. atmospheric pollution

B. boating

C. dumped waste

D. land runoff

3 The students in Brian's school can take up to three electives each year. The Venn diagram below shows the electives chosen by students in Brian's class.

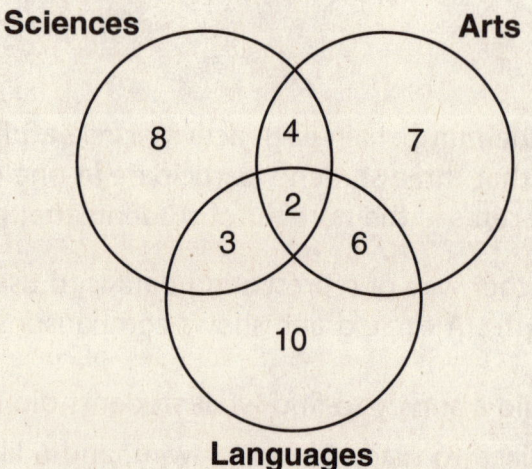

How many students take languages as an elective?

A. 10

B. 21

C. 15

D. 11

4 400 students from all over the country attended a technology summer camp in Boston. The circle graph below shows the percent of students from the different regions of the country.

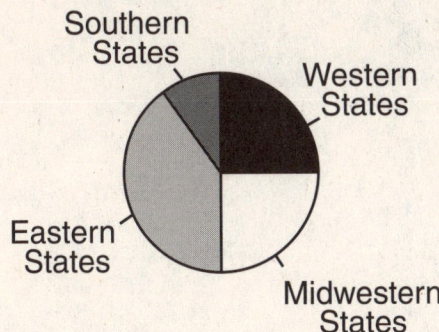

From which region did 160 students attend?

A. Eastern states

B. Midwestern states

C. Southern states

D. Western states

5 The elements in the human body are given in the table below.

Element	Percent
Carbon	20
Oxygen	70
Other	10

a. Calculate the number of degrees that would be in each sector of the circle graph for the data above. Show your work.

b. Sketch a circle graph showing the data in the table.

Answer Key
for Miles

Mile 4

1. −2.4, no
2. 9.14, yes
3. −8, yes
4. 2.2, no
5. 4.6$\overline{6}$, yes
6. 0.3$\overline{3}$, yes
7. 7.5, no
8. −5, no

The answer to the riddle is A PELICAN.

---

1. C
2. A
3. D
4. C
5. B
6. C
7. A
8. D
9. B
10. D

Mile 5

1. M Commutative property of addition
2. A Inverse property of multiplication
3. T Distributive property
4. T Commutative property of multiplication
5. E Identity property of multiplication
6. R Associative property of addition
7. H Inverse property of addition
8. O Identity property of addition
9. R Associative property of multiplication
10. N Multiplicative property of zero

The answer is MATTERHORN.

---

1. D
2. C
3. B
4. A
5. B
6. C
7. C
8. B
9. D
10. A
11. B

Mile 6

1. B

2. A

3. This question has three parts.

a. 2 quarters

   1 quarter, 2 dimes, and 1 nickel

   1 quarter, 1 dime, and 3 nickels

   3 dimes and 4 nickels

b. No; she would need 3 quarters and 3 pennies to make $0.78 with 6 coins. She does not have 3 quarters.

c. 5. She can use 1 quarter, 1 dime, 1 nickel and 2 pennies. That is the combination of the least number of coins to have a total value of 42 cents.

Mile 7

1. B

2. C

3. This question has two parts.

a. 133

b. 81. The digits of Jean's favorite number add up to 9, and the number must be an odd multiple of 9 that is less than 100. There are only four possible numbers: 27, 45, 63, and 81. The number 27 has 4 factors: 1, 3, 9, and 27. The number 45 has 6 factors: 1, 3, 5, 9, 15, and 45. The number 63 has 6 factors: 1, 3, 7, 9, 21, and 63. The number 81 has 5 factors: 1, 3, 9, 27, and 81, and fits Jean's description of her favorite number.

Mile 8

1. W 9 to the sixth power

2. O y to the third power

3. R 5^2

4. C $y \times y \times y \times y$

5. E $5 \times y \times y \times y \times y$

6. S $2 \times 2 \times y \times y$

7. T 6 to the ninth power

8. E $3 \times 4 \times y \times y$

9. R $\frac{1}{125}$

10. C $y^5 \times y^2$

11. E 9^8

12. N $\frac{1}{36}$

13. T $2y^2$

14. R $4 \times 4 \times 4$

15. U $6 \times y \times y \times y$

16. M 256

The answer is WORCESTER CENTRUM.

---

1. C	10. A
2. C	11. $\frac{1}{8}$
3. A	12. $\frac{1}{25}$
4. B	13. 3^7
5. C	
6. A	
7. 64	
8. A	
9. 36	

Mile 9

1. 40 = A_
2. 12 = O
3. $\frac{16}{21}$ = W
4. 19.4 = N
5. $10\frac{8}{9}$ = R

A	N		A	R	R	O	W	!
40	19.4		40	13	13	12	$\frac{16}{21}$	

1. B
2. D
3. A
4. C
5. C
6. A
7. D
8. C
9. −19
10. 16
11. 28
12. −41
13. 6
14. 2

Mile 10

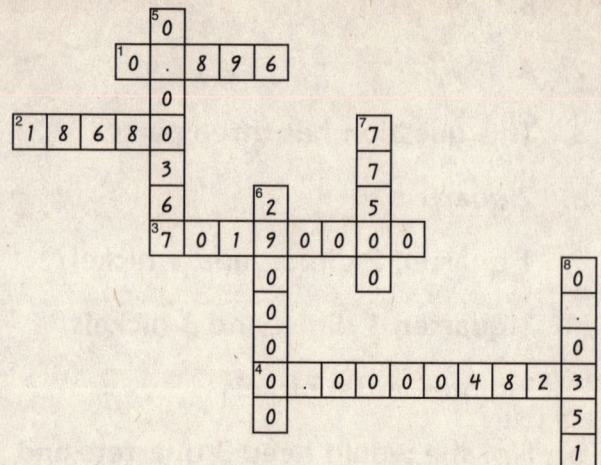

1. 8.436×10^{-3}
2. B
3. D
4. 1.34×10^5
5. 8.01×10^8
6. B
7. A
8. D
9. 5.3446×10^{-3}
10. B
11. A

Mile 11

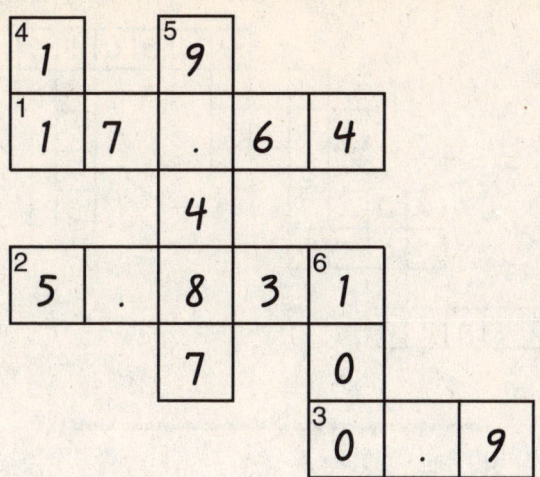

1. C
2. A
3. D
4. B
5. D
6. D
7. A
8. D
9. B
10. B
11. A

Mile 12

1. A $1\frac{7}{8}$
2. B $1\frac{1}{8}$
3. O $1\frac{1}{4}$
4. N $1\frac{1}{2}$
5. U $1\frac{5}{8}$
6. S $1\frac{3}{8}$

1		2	3	4	5	6	
A		B	O	N	U	S	!

1. B
2. B
3. C
4. B
5. D
6. A
7. D
8. A
9. B
10. B
11. C

Mile 13

1. B
2. A
3. B
4. C
5. C

Mile 14

1. 12%
2. 9
3. 12
4. 40%
5. 7
6. $33\frac{1}{3}$%
7. 6
8. 3.3
9. 3.5
10. 225%

---

1. B
2. C
3. A
4. D
5. 49
6. 180
7. 60
8. 25
9. 30
10. A
11. D
12. C

Mile 15

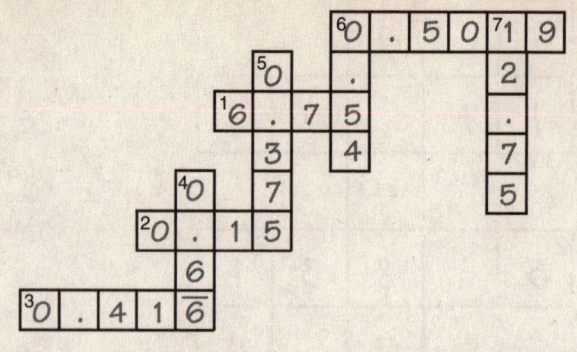

---

1. B
2. C
3. D
4. A
5. C
6. B
7. A
8. B
9. A
10. D
11. A

Mile 16

1. D 1
2. O 20
3. G 9
4. W 4
5. O 2
6. O 15
7. D 6

The answer is DOGWOOD.

---

1. D
2. B
3. C
4. A
5. A
6. D
7. B
8. C
9. B
10. C
11. A
12. B

Mile 17

1. $\frac{2}{3}$
2. 25
3. 10
4. $\frac{3}{7}$
5. $\frac{10}{3}$
6. 14
7. 28
8. 15
9. 40
10. 1.5

I	T		W	A	S		H	O	T	!
$\frac{10}{3}$	25		15	$\frac{2}{3}$	1.5		14	$\frac{3}{7}$	25	

---

1. C
2. D
3. B
4. A
5. B
6. D
7. $\frac{4}{3}$
8. $\frac{P}{s} = 4$
9. $\frac{V}{h} = B$
10. A

```
 ¹4
 ²6 ³1
    ⁴4  4  ⁵8        ⁶2
    5      7          2
    8      1          0
        ⁷3  4   0  0
                     0
```

1. A

2. A

3. B

4. B

5. C

6. D

7. C

8. C

You should draw a triangle with 6 dots on each side for a total of 21 dots.

1. 30

2. This question has 3 parts.

a. 10

b.

Drawing	Number of White Hexagons
1	9
2	12
3	15

c. The rule is $3n + 6$. Based on the table in part *b*, you can see that the numbers of white hexagons are increasing multiples of 3. If you add 2 to the drawing number and multiply by 3, $(n + 2) \times 3$, you will get the pattern. The expression $(n + 2) \times 3$ simplifies to $3n + 6$.

Mile 20

1. 21
2. 4
3. 9
4. 16
5. −1
6. 36
7. −10
8. 27
9. 1
10. 7
11. 11
12. −5
13. 25
14. 8

---

1. A
2. −16
3. 13
4. 21
5. B
6. A
7. (1, 4) or (3, 6)
8. D
9. A
10. D
11. D
12. B

Mile 21

1. $10 + c = 115$
2. $p - 5 < 50$
3. $t \times a = s$
4. $\frac{m}{c} + 30 = j$
5. $2l - 7 \geq 14$

Note: You may use any variables for these sentences.

---

1. D
2. C
3. A
4. B
5. A
6. A
7. D
8. B
9. C
10. A
11. D

¹2	1	
0		

²1	4	³2
6	■	1
⁴9	0	2

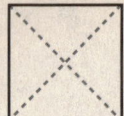 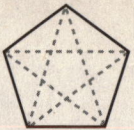 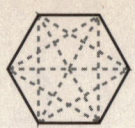 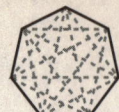

Number of Sides (n)	Number of Diagnals (d)
4	2
5	5
6	9
7	14

---

1. B

2. C

3. 9

4. 3

5. −1

6. 4

7. B

8. A

9. B

10. A

11. D

---

1. A

2. B

3. C

4. A

5. B

6. D

7. A

8. B

Mile 24

The following are several different ordered pairs that represent solutions to the linear equation $y = 2x - 3$:

(−4, −11)	(2, 1)
(−3, −9)	(3, 3)
(−2, −7)	(4, 5)
(−1, −5)	(5, 7)
(0, −3)	(6, 9)
(1, −1)	(7, 11)

The answers above are not the only ones that are correct. However, they are the only ones that could fit on the grid provided.

1. B

2. C

3. A

4. B

5. D

6. B

7. A

Mile 25

1. B

2. D

3. B

4. A

5. A

Mile 26

1. QUADRILATERAL

2. PENTAGON

3. RIGHT

4. POLYGON

5. TRAPEZOID

6. NONAGON

7. RHOMBUS

8. EQUILATERAL

1. This question has four parts.

a. The number of triangles is the number of sides minus 2, or $n - 2$.

b. 1,260°

c. Subtract 2 from the number of sides in the polygon, and then multiply by 180°, or $(n - 2)180$.

d. 12

2. D

3. C

4. D

5. C

6. A

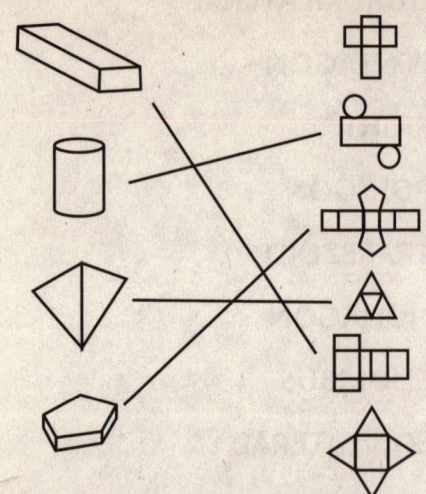

1. D

2. C

3. B

4. C

5. A

6. This question has four parts.

a. 10 faces, 16 vertices, 24 edges

b. $2f - 4 = v$

c. 9 faces, 9 edges, 16 edges

d. $2f - 2 = e$

1. F

2. T

3. T

4. F

5. T

6. F

7. F

8. T

---

1. B

2. C

3. A

4. B

5. D

6. A

7. C

Mile 29

1. C
2. A

Mile 30

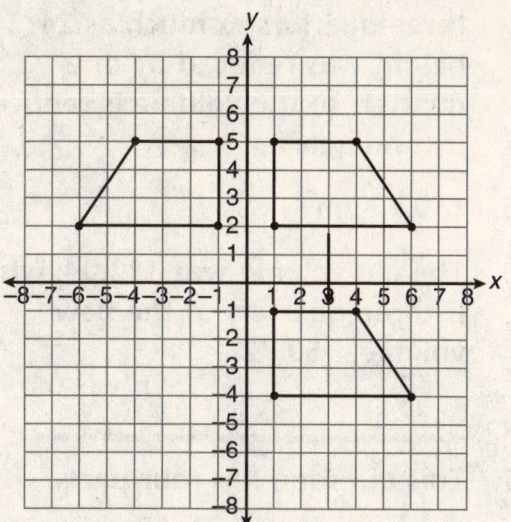

1. C
2. D
3. B
4. B
5. C
6. D
7. D
8. A

Mile 31

1. 9.2 mi.
2. 11.4 mi.
3. 15.6 mi.
4. 12 mi.
5. The total distance they traveled is 48.2 mi.

---

1. A
2. D
3. B
4. C
5. D
6. B
7. C
8. C

Mile 32

1. A
2. C
3. D
4. B
5. A
6. B
7. C
8. D
9. C
10. A

Mile 33

1. C
2. A
3. B
4. C
5. D
6. B

Mile 34

1. $84
2. D
3. 50.24 ft.^{2}
4. A
5. C

Mile 35

1. 24 in.^{3}
2. 113.04 cm^3
3. 18 in.^{3}
4. The new volume is 18, which is three-quarters as much as 24. The height was reduced by one-quarter, so the volume is reduced by one-quarter.
5. 150.72 cm^3
6. The old volume was 113.04, which is three-quarters of the new volume, 150.72.

---

1. This question has four parts.
   a. 4 cm
   b. $\pi r^2 h = (3.14)(4^2)(10) = 502.4$ cm^3
   c. $\pi r^2 h = (3.14)(2^2)(5) = 62.8$ cm^3
   d. 8:1
2. C
3. A
4. D
5. C

Mile 36

```
 1      2
┌──┬──┬──┐
│1 │1 │9 │
├──┼──┼──┤
│0 │  │
├──┼──┼──┬──┐   3
│4 │  │  │1 │
├──┼──┼──┼──┼──┬──┬──┐   4        5
│5 │0 │3 │  │1 │6 │
├──┼──┼──┼──┼──┼──┤
│3 │  │  │7 │
├──┼──┼──┼──┤   6
│6 │1 │1 │
└──┴──┴──┘
```

1. C
2. B
3. A
4. B
5. C
6. D
7. A
8. C
9. B

Mile 37

Possible combinations:

Pizza with no toppings

Pizza with sausage

Pizza with peppers

Pizza with extra cheese

Pizza with sausage and peppers

Pizza with sausage and extra cheese

Pizza with peppers and extra cheese

Pizza with sausage, peppers, and extra cheese

There are 8 possible combinations to choose from.

1. C

2. 3, 2, 2; 3, 2, 4; 3, 2, 6; 3, 4, 2; 3, 4, 4; 3, 4, 6; 3, 6, 2; 3, 6, 4; 3, 6, 6

3. $\frac{1}{9}$

4. D

Mile 38

1. $\frac{3}{10}$ or 0.30

2. $\frac{1}{365}$ or 0.00273

3. $\frac{10}{20} = \frac{1}{2}$ or 0.50

4. $\frac{3}{12} = \frac{1}{4}$ or 0.25

5. $\frac{3}{30} = \frac{1}{10}$ or 0.10

6. $\frac{4}{50} = \frac{2}{25}$ or 0.08

7. $\frac{3}{25}$ or 0.12

8. $\frac{28}{365}$ or 0.077

9. $\frac{18}{34}$ or 0.53

10. $\frac{14}{60}$ or 0.233

---

1. 12

2. 10

3. C

4. B

5. C

6. $\frac{1}{3}$

7. This question has two parts.

a. $1 \times 1 = 1$

$2 \times 1 = 2$

$3 \times 1 = 3$

$2 \times 2 = 4$

$3 \times 2 = 6$

$3 \times 3 = 9$

b. $\frac{5}{8}$. If Jerry draws an orange 3 card, there are 8 more cards remaining: orange 1, orange 2, red 1, red 2, red 3, blue 1, blue 2, and blue 3. He will only score points if he draws an orange or red card. There are 5 red or orange cards out of the 8 remaining cards, so the probability that he will score points is $\frac{5}{8}$.

Mile 39

1. D

2. A

3. B

4. A

5. This question has two parts.

a. Carbon: $\frac{20}{100} = \frac{72}{360}$ or 72°

Oxygen: $\frac{70}{100} = \frac{252}{360}$ or 252°

Other: $\frac{10}{100} = \frac{36}{360}$ or 36°

b.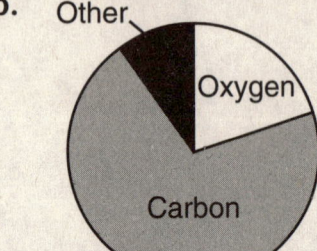

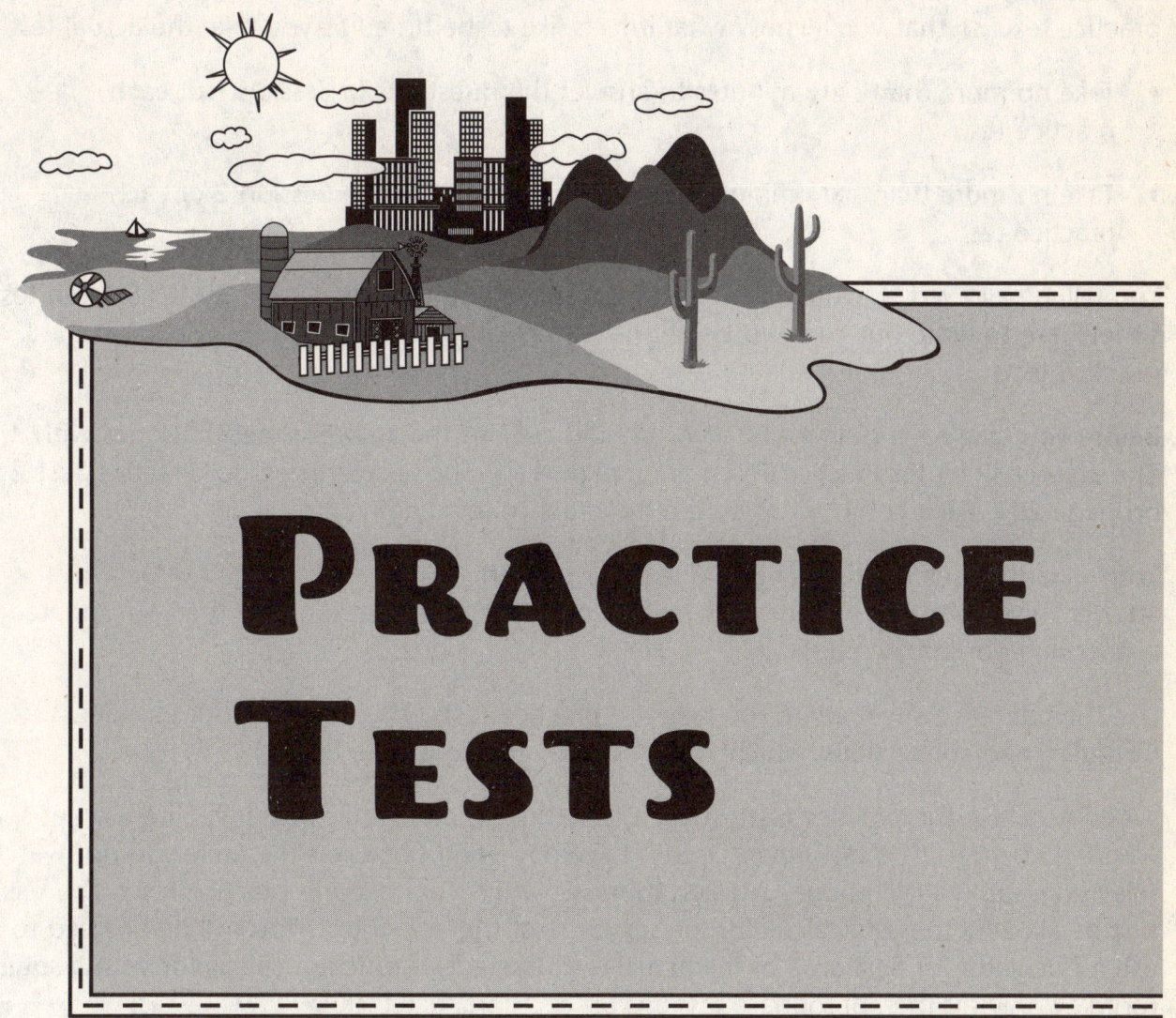

PRACTICE TESTS

HOW TO TAKE THE PRACTICE TESTS

Try to take each practice test as if it were the actual Grade 8 MCAS Math test. Take both sessions of one practice test on the same day with a break in between sessions. If you begin a session, don't stop until you are finished. You should time yourself while you take the practice tests so that you'll know what it feels like to be timed as you take the actual test.

- Take no more than sixty minutes to answer the questions in Session 1 of each practice test.

- Take no more than sixty minutes to answer the questions in Session 2 of each practice test.

You will be allowed to write on the test book while you take the Grade 8 MCAS Math test, so feel free to write out your work right next to each question that you answer on the practice tests.

Before you take each practice test, you should cut out the answer sheet that goes with it. The answer sheet for Practice Test 1 is on page 163. The answer sheet for Practice Test 2 is on page 201. Each comes right before the practice test begins.

You will answer all multiple-choice questions by filling in the appropriate bubbles on the answer sheets. Answer all short-answer and open-response questions in the appropriate spaces on your practice tests.

Use the reference sheet while you take the practice tests. This sheet includes a ruler, formulas, and conversions, which you may need to use on the test.

When you take the practice tests, make the setting as realistic as possible. Find a quiet spot where you will not be disturbed. Do not have any open books on the table. Do not watch television, talk on the phone, or listen to music when you take the practice tests. You will **not** be allowed to use a calculator for Session 1 of the actual test. You will be allowed to use a calculator for Session 2 of the actual test. Use a four function calculator with a square root key only for the appropriate sessions on each practice test.

Remember to use the tips and skills that you have learned and practiced in this book. They will help you do your best on the practice tests. After you have taken each practice test, go over it with an adult. The answers and explanations to the practice tests begin on page 237. Read the explanations of the correct answers for as many questions as you need to. Pay special attention to the explanations of questions that you answered incorrectly or had trouble answering.

Good luck!

Practice Test 1

PRACTICE TEST 1

Session 1

1. Ⓐ Ⓑ Ⓒ Ⓓ
2. Ⓐ Ⓑ Ⓒ Ⓓ
3. Ⓐ Ⓑ Ⓒ Ⓓ
4. Ⓐ Ⓑ Ⓒ Ⓓ
5. Ⓐ Ⓑ Ⓒ Ⓓ
6. Ⓐ Ⓑ Ⓒ Ⓓ
7. Ⓐ Ⓑ Ⓒ Ⓓ
8. Ⓐ Ⓑ Ⓒ Ⓓ
9. Ⓐ Ⓑ Ⓒ Ⓓ
10. Ⓐ Ⓑ Ⓒ Ⓓ
11. Use space provided.
12. Use space provided.
13. Use space provided.
14. Use space provided.
15. Use space provided.
16. Ⓐ Ⓑ Ⓒ Ⓓ
17. Ⓐ Ⓑ Ⓒ Ⓓ
18. Ⓐ Ⓑ Ⓒ Ⓓ
19. Ⓐ Ⓑ Ⓒ Ⓓ
20. Ⓐ Ⓑ Ⓒ Ⓓ
21. Ⓐ Ⓑ Ⓒ Ⓓ
22. Ⓐ Ⓑ Ⓒ Ⓓ
23. Ⓐ Ⓑ Ⓒ Ⓓ
24. Ⓐ Ⓑ Ⓒ Ⓓ
25. Use space provided.
26. Use space provided.
27. Use space provided.
28. Use space provided.

Session 2

29. Ⓐ Ⓑ Ⓒ Ⓓ
30. Ⓐ Ⓑ Ⓒ Ⓓ
31. Ⓐ Ⓑ Ⓒ Ⓓ
32. Ⓐ Ⓑ Ⓒ Ⓓ
33. Ⓐ Ⓑ Ⓒ Ⓓ
34. Ⓐ Ⓑ Ⓒ Ⓓ
35. Ⓐ Ⓑ Ⓒ Ⓓ
36. Use space provided.
37. Use space provided.
38. Ⓐ Ⓑ Ⓒ Ⓓ
39. Ⓐ Ⓑ Ⓒ Ⓓ
40. Ⓐ Ⓑ Ⓒ Ⓓ
41. Ⓐ Ⓑ Ⓒ Ⓓ
42. Ⓐ Ⓑ Ⓒ Ⓓ
43. Ⓐ Ⓑ Ⓒ Ⓓ
44. Ⓐ Ⓑ Ⓒ Ⓓ
45. Ⓐ Ⓑ Ⓒ Ⓓ
46. Ⓐ Ⓑ Ⓒ Ⓓ
47. Ⓐ Ⓑ Ⓒ Ⓓ
48. Use space provided.

MATHEMATICS REFERENCE SHEET

PERIMETER FORMULAS

square.........$P = 4s$

rectangle.........$P = 2b + 2h$

triangle.........$P = a + b + c$

Pythagorean theorem

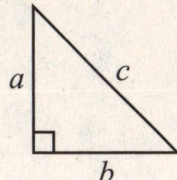

$a^2 + b^2 = c^2$

CIRCUMFERENCE FORMULAS

circle.........$C = 2\pi r$

OR

$C = \pi d$

CONVERSIONS

1 mile = 5,280 feet

1 square mile = 640 acres

AREA FORMULAS

square.........$A = s^2$

rectangle.........$A = bh$

OR

$A = lw$

triangle.........$A = \frac{1}{2}bh$

circle.........$A = \pi r^2$

trapezoid.........$A = \frac{1}{2}h(b_1 + b_2)$

VOLUME FORMULAS

rectangular prism......$V = Bh$
(B = area of base)

cone.....................$V = \frac{1}{3}\pi r^2 h$

cylinder................$V = \pi r^2 h$

cube.....................$V = s^3$
(s = length of an edge)

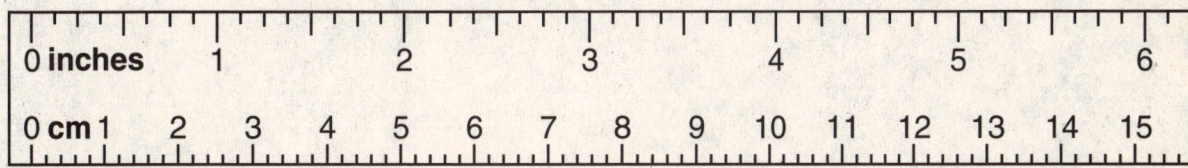

Session 1

1 Which of the following is the best unit in which to measure the area of a sheet of notebook paper?

   A. cubic centimeters

   B. liters

   C. millimeters

   D. square centimeters

2 What is the value of $4 + (-12) \div 2 - 2^3$?

   A. -12

   B. -10

   C. $\;\;\;4$

   D. $\;\;\;6$

3 Jared is helping his mother tile the bathroom floor. They measured the area of the floor to be 92 square feet. Jared and his mother purchased enough tiles to cover the floor and an extra 10% to allow for mistakes. About how much tile did they purchase?

   A. 10 square feet

   B. 40 square feet

   C. 95 square feet

   D. 100 square feet

4 The graph below represents the coordinates for which table of values?

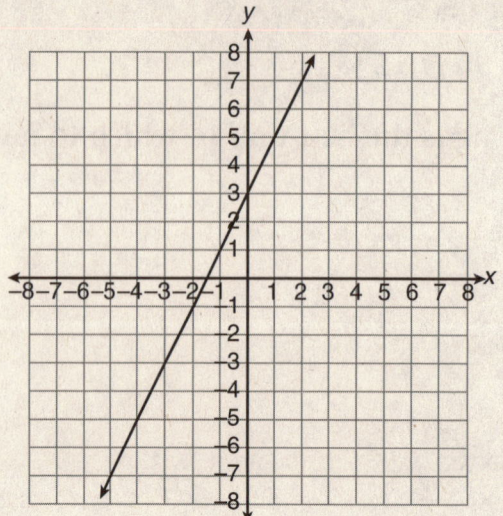

A.

x	y
−1	1
0	3
1	5

B.

x	y
−1	−2
0	0
1	2

C.

x	y
−1	1
0	2
1	3

D.

x	y
0	−1
3	0
5	1

5 Which of the following numbers have a square root that is about 7.5?

    A. 40

    B. 56

    C. 66

    D. 70

6 Which of the following has a square root that is between 6 and 7?

    A. 38

    B. 52

    C. 70

    D. 92

7 Karrie spins the spinner below and flips a coin at the same time. Which of the following represents the set of possible outcomes? *H* stands for heads and T stands for tails.

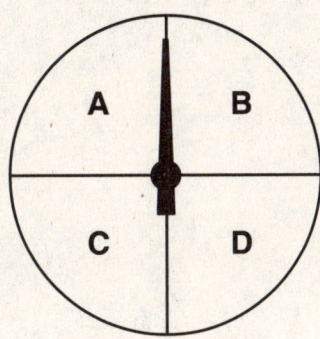

    A. {A, H; B, H; C, H; D, H}

    B. {A, T; B, T; C, T; D, T}

    C. {A, H; B, H; C, H; A, T; B, T; C, T}

    D. {A, H; B, H; C, H; D, H; A, T; B, T; C, T; D, T}

8 The sixth term of a sequence is 46. To which of the following sequences does the term belong?

A. 1, 4, 7, 10, _____, _____

B. 1, 4, 10, 19, _____, _____

C. 1, 4, 13, 40, _____, _____

D. 1, 4, 16, 64, _____, _____

9 Shannon had $25. She purchased a shirt for $15 and bought 3 pairs of socks. If the pairs of socks all cost the same amount of money, which inequality can be used to find the greatest amount that Shannon spent for each pair of socks?

A. $3s + \$15 \geq \$25$

B. $3s + \$25 \leq \$15$

C. $3s + \$15 \leq \$25$

D. $3s + \$25 \geq \$15$

10 If $a = 4$, and $b = -12$, what is the value of $5 + [b + a(3 - a)]$?

A. −11

B. −3

C. 3

D. 13

11 Fiona trimmed a tree in her backyard that was 20 feet tall. She cut 15% off its height. How tall was the tree after Fiona trimmed it?

12 What is the value of -6^2?

13 Beyonce has gray socks, white socks, and blue socks in her drawer. If she removes a sock from the drawer, the probability that it is gray is $\frac{4}{9}$. The probability that the sock is white is $\frac{2}{9}$. The probability that the sock is blue is $\frac{1}{3}$. If there are 3 blue socks in the drawer, how many are gray?

14 Kendall folded the square pyramid with the measurements shown from the pattern below. The dotted lines indicate where the pattern was folded.

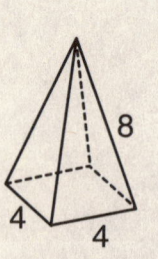

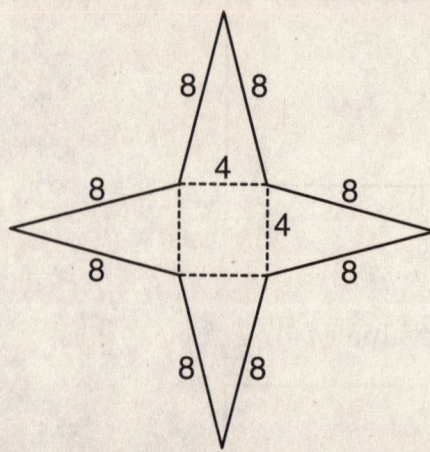

a. Kendall wants to fold a triangular pyramid. Sketch a pattern for a triangular pyramid, showing fold lines, and label a possible length for each line segment.

b. Draw a net that can be folded into a cone.

c. If the height of your cone is 8 units and the radius of the base is 6 units, what would be the circumference of the circular base of the cone? Explain your answer.

15 During their show, the marching band played a number of songs. Each time a new song was played, a new group of musicians joined the rest. The pattern is shown in the diagram below.

Song 1   Song 2   Song 3

a. If the pattern continued, how many new musicians would have joined in for the fourth song?

b. If the pattern continued, how many musicians would have there been altogether for the fifth song?

c. Make a table showing the song number and the number of musicians for the first six songs.

d. What is the relationship between the song number and the number of musicians?

16 In a local elementary school, there are 200 computers. The computers are distributed among five grade levels as follows:

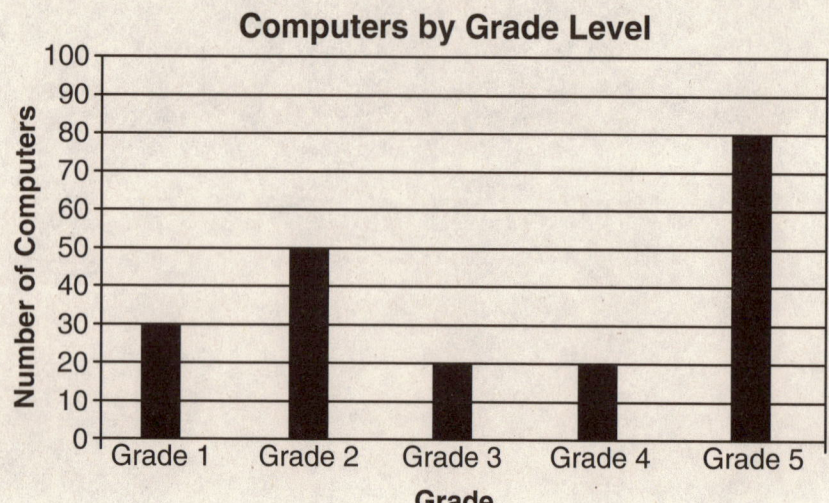

Which two grade levels combined have the same percentage of computers as the fifth grade?

A. grades 1 and 2

B. grades 1 and 3

C. grades 2 and 3

D. grades 3 and 4

17 What is the length of the longest side of the right triangle shown in the diagram below?

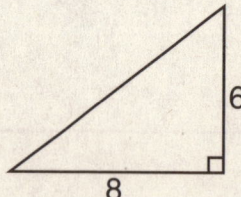

A. 8 units

B. $8\frac{1}{2}$ units

C. 9 units

D. 10 units

Roadmap to the MCAS: Grade 8 Math

18 If the rectangle below is reflected across the *y*-axis and then shifted 4 units to the left, which of the following are the coordinates of its image?

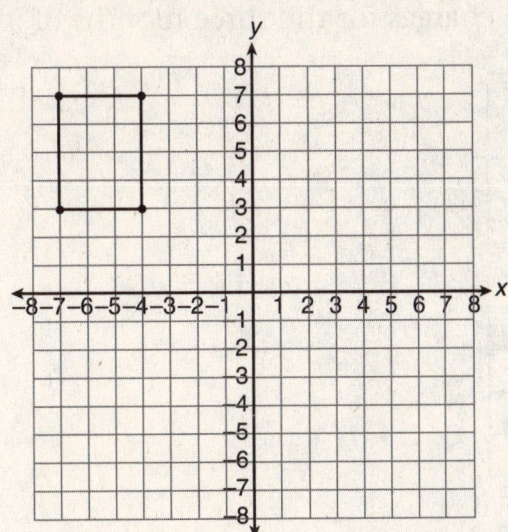

A. (–7, –3), (–7, –7), (–4, –7), (–4, –3)

B. (–4, 3), (–4, –6), (0, 3), (0, 6)

C. (0, 3), (0, 7), (3, 7), (3, 3)

D. (4, 3), (4, 7), (7, 7), (7, 3)

19 A long-distance telephone company charges a $4.50 monthly fee, plus 15 cents for each minute. Geri used 10 minutes of long-distance service the first month, 20 minutes the second month, and 30 minutes the third month. Which of the following shows the charges for the three months of long-distance calls?

A.

Month	Cost
1	$1.50
2	$3.00
3	$4.50

B.

Month	Cost
1	$4.50
2	$9.00
3	$13.50

C.

Month	Cost
1	$6.00
2	$7.50
3	$9.00

D.

Month	Cost
1	$6.75
2	$9.00
3	$15.00

20 What is the value of (0.3)(0.3)(0.3)(0.3)?

    A. 0.00081

    B. 0.0081

    C. 0.081

    D. 0.81

21 The letters for the word *SPELLING* are placed in a hat. One letter is chosen at random. What is the probability that the letter is NOT a vowel?

    A. $\frac{1}{4}$

    B. $\frac{1}{3}$

    C. $\frac{1}{2}$

    D. $\frac{3}{4}$

22 12 out of 25 students in a class ride their bikes to school. What percent of the students ride their bikes to school?

    A. 40%

    B. 48%

    C. 50%

    D. 52%

23 An experiment was done to record the growth of a certain plant according to the number of days of sunshine the plant received. The results are shown in the table.

Plant Growth (g)	Days of Sunshine (s)
5	12
8	18
11	22
14	30

Which of the following equations shows the approximate relationship between the number of days of sunshine, s, and the growth of the plant, g?

A. $s = \frac{1}{2}g$

B. $s = 2g$

C. $s = g + 15$

D. $s = g - 15$

24 Nala is traveling 206 miles to visit her cousins. She has already traveled $\frac{1}{4}$ of the way. About how far does Nala still have to travel?

A. 50 miles

B. 80 miles

C. 100 miles

D. 150 miles

25 What is the ratio of the diameter of a circle to its radius?

26 What is a possible ordered pair for the equation $2a + 3 = b$, if a is an odd number greater than 5 and less than 10? Write the ordered pair in the form (a, b).

27 Ms. Froome wants to put sod in a rectangular area that measures 84 inches by 96 inches. The sod she will use costs $15.50 per square foot. How much will it cost Ms. Froome to sod this area?

28 A game is being played using the numbered cards below. The cards are placed facedown and a player selects two cards. If the sum of the numbers on the cards is 10 or greater, player A wins. If the sum of the numbers on the cards is less than 10, player B wins.

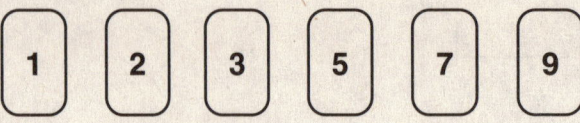

a. Make a list of the possible outcomes for the sums when two cards are selected.

b. If the first card is a 5, what is the probability that player B will win?

c. Is the game fair? Explain your answer.

29 Which decimal has the same value as $\frac{2}{5} \div \frac{1}{3}$?

A. 0.13

B. 0.20

C. 0.83

D. 1.20

30 A car rental company charges its customers depending on the number of miles driven. The company uses the chart below to show its rates.

Distance	Cost
First 100 miles	$150
101–200 miles	$190
201–300 miles	$230
301–400 miles	$270

According to the pattern in the chart, what is the rate if a customer travels between 801 and 900 miles?

A. $390

B. $430

C. $470

D. $550

31 Kris charges $5 an hour to tutor math students. If he works from 20 to 30 hours each month, which of the following shows the amount, *x*, he earns in one month?

A.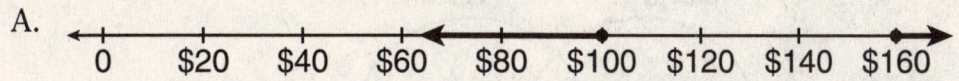

B.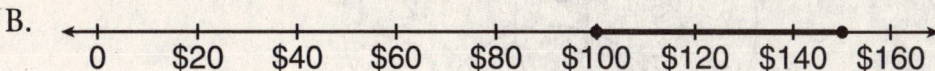

C.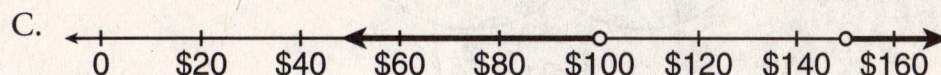

D.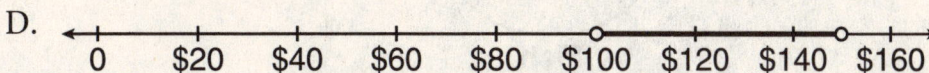

32 The Venn diagram below shows the number of pets that live in homes in a neighborhood.

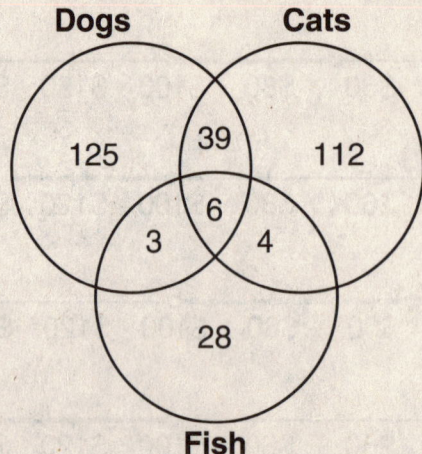

How many households have fish as pets?

A. 13

B. 28

C. 38

D. 41

33 Which of the following CANNOT be true in the equation $a = \frac{1}{b} - 4$?

A. b is 0.

B. b is negative.

C. b is greater than a.

D. b and a are both negative.

34 Which of the following has the same value as $\frac{3}{x}$?

A. $3 \div \frac{1}{x}$

B. $3 \times \frac{1}{x}$

C. $3x$

D. $x \div \frac{1}{3}$

35 In the diagram below, *XY* and *UV* are parallel lines. What is the measure of the angle marked *m*?

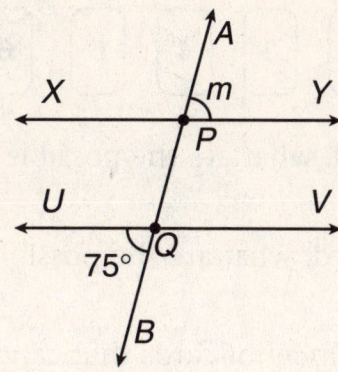

A. 75°

B. 90°

C. 100°

D. 105°

36 Kiara played a game using the cards below. In this game, the sum of the cards selected must equal 10 to score a point.

a. If two cards are selected, what are the possible combinations that will score the player a point?

b. If three cards are selected, what are the possible combinations that will score the player a point?

c. What is the greatest number of cards that can be selected to get a sum of 10? Explain your answer.

37 A cone and a cylinder both have the same height and base. The height is 8 in. The base is shown in the diagram below.

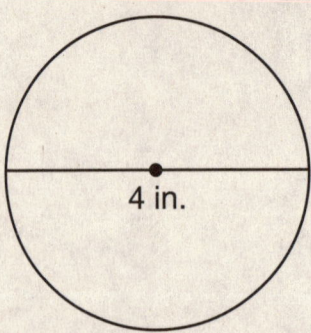

4 in.

a. How much water will fit in the cone? Round your answer to the nearest hundredth. (Use your reference sheet to help you find the formula.)

b. How much water will fit in the cylinder? Round your answer to the nearest hundredth. (Use your reference sheet to help you find the formula.)

c. What is the ratio of the volume of the cone to the volume of the cylinder? Explain how the formulas could help you determine the ratio.

38 Six different stores carry the same model of a portable CD player. The prices at the different stores are as follows:

$20.00, $45.00, $15.00, $30.00, $25.00, $15.00

What is the mode of the prices?

A. $15.00

B. $22.50

C. $25.00

D. $30.00

39 The school store has the following school sweatshirts in stock:

Color	Style	Quantity
gray	crewneck	3
black	hooded	4
white	crewneck	8
red	hooded	4
blue	crewneck	12
green	hooded	3

If a student purchased a sweatshirt, what is the probability that the student purchased a black hooded sweatshirt?

A. 12%

B. 20%

C. 35%

D. 60%

40 Selena rented a carpet cleaner to clean her carpets. The rental will cost between $30 and $40. She plans to keep the machine for 6 hours. Which of the following inequalities shows the hourly cost, c, for the machine?

A. $\$30 \leq 6c \leq \$40$

B. $\$30 \geq 6c \geq \$40$

C. $\$30 \leq 6c \geq \$40$

D. $\$30 \geq 6c \leq \$40$

41 The table of values shown below satisfies which of the following equations?

x	y
-2	-5
-1	-3
0	-1
1	1

A. $y = x - 3$

B. $y = x + 3$

C. $y = 2x - 1$

D. $y = 2x + 1$

42 What is y in the equation $5y = 9 + 3(y + 5)$?

    A. 2

    B. 3

    C. 8

    D. 12

43 For the equation $2ab = -\frac{1}{2}$, which statement CANNOT be true?

    A. a is greater than 1.

    B. b is less than 1.

    C. Either a or b is 0.

    D. Either a or b is negative.

44 Anita has a piano recital at 6:30 P.M. She has to travel 35 miles to get to the recital. She takes a train that travels 80 miles per hour. At what time should she leave to arrive closest to her recital time?

    A. 5:00 P.M.

    B. 5:30 P.M.

    C. 6:00 P.M.

    D. 6:30 P.M.

45 Montel's class raised $75 to go on a school trip to a local museum. The class visited 3 different exhibits and purchased souvenirs. Each exhibit cost $20 for admission for the entire class. If the class had $4 left over at the end of the day, which of the following equations could be used to find the cost of the souvenirs, s?

    A. $75 = 3 \times 20 + s + 4$

    B. $75 = 3s + 20 + 4$

    C. $75 = \frac{20}{3} + s + 4$

    D. $75 = 3s + 20 - 4$

46 Triangle *XYZ* below is translated so that the image of point *X* moves to (−1, 0). What are the coordinates for the image of point *Z*?

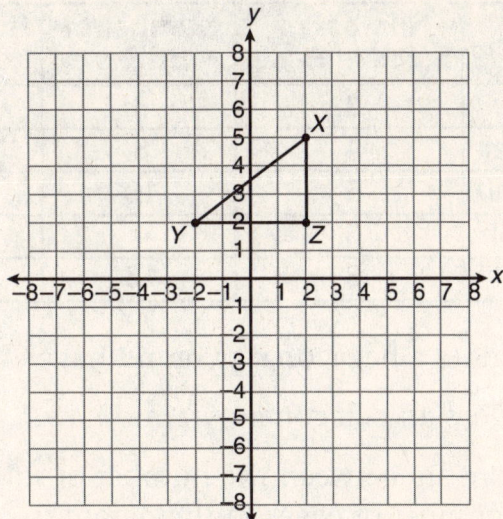

A. (−5, −3)

B. (−1, −3)

C. (1, 3)

D. (5, 3)

47 During lunch, $\frac{2}{5}$ of the students in the cafeteria had juice with their meal and $\frac{1}{3}$ had milk. The remaining 24 students had water. How many students were in the cafeteria at lunch?

A. 44

B. 90

C. 100

D. 114

48 Tara made a table showing the number of edges and the number of faces in some common pyramids. Tara's table is shown below.

Pyramid	Number of Sides in Base	Number of Edges	Number of Faces
Triangular	3	6	4
Rectangular	4	8	5
Pentagonal	5	10	6
Hexagonal	6		7
Octagonal	8	16	

a. How many edges does a hexagonal pyramid have?

b. How many faces does an octagonal pyramid have?

c. Explain the relationship between the number of sides in the base of a pyramid and the number of edges in the pyramid.

d. Explain the relationship between the number of sides in the base of a pyramid and the number of faces of the pyramid.

PRACTICE TEST 2

PRACTICE TEST 2

Session 1

1. (A) (B) (C) (D)
2. (A) (B) (C) (D)
3. (A) (B) (C) (D)
4. (A) (B) (C) (D)
5. (A) (B) (C) (D)
6. (A) (B) (C) (D)
7. (A) (B) (C) (D)
8. Use space provided.
9. Use space provided.
10. Use space provided.
11. (A) (B) (C) (D)
12. (A) (B) (C) (D)
13. (A) (B) (C) (D)
14. Use space provided.
15. (A) (B) (C) (D)
16. (A) (B) (C) (D)
17. (A) (B) (C) (D)
18. (A) (B) (C) (D)
19. (A) (B) (C) (D)
20. (A) (B) (C) (D)
21. (A) (B) (C) (D)
22. (A) (B) (C) (D)
23. Use space provided.
24. Use space provided.
25. Use space provided.
26. Use space provided.
27. Use space provided.

Session 2

28. (A) (B) (C) (D)
29. (A) (B) (C) (D)
30. (A) (B) (C) (D)
31. (A) (B) (C) (D)
32. (A) (B) (C) (D)
33. (A) (B) (C) (D)
34. (A) (B) (C) (D)
35. (A) (B) (C) (D)
36. Use space provided.
37. Use space provided.
38. (A) (B) (C) (D)
39. (A) (B) (C) (D)
40. (A) (B) (C) (D)
41. (A) (B) (C) (D)
42. (A) (B) (C) (D)
43. (A) (B) (C) (D)
44. (A) (B) (C) (D)
45. (A) (B) (C) (D)
46. (A) (B) (C) (D)
47. (A) (B) (C) (D)
48. Use space provided.

MATHEMATICS REFERENCE SHEET

PERIMETER FORMULAS

square.........$P = 4s$

rectangle.........$P = 2b + 2h$

triangle.........$P = a + b + c$

Pythagorean theorem

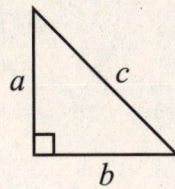

$$a^2 + b^2 = c^2$$

CIRCUMFERENCE FORMULAS

circle.........$C = 2\pi r$

OR

$C = \pi d$

CONVERSIONS

1 mile = 5,280 feet

1 square mile = 640 acres

AREA FORMULAS

square.........$A = s^2$

rectangle.........$A = bh$

OR

$A = lw$

triangle.........$A = \frac{1}{2}bh$

circle.........$A = \pi r^2$

trapezoid.........$A = \frac{1}{2}h(b_1 + b_2)$

VOLUME FORMULAS

rectangular prism......$V = Bh$
(B = area of base)

cone....................$V = \frac{1}{3}\pi r^2 h$

cylinder.................$V = \pi r^2 h$

cube....................$V = s^3$
(s = length of an edge)

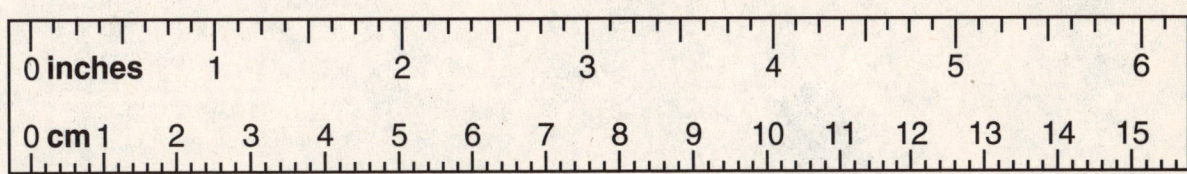

Session 1

1 What is the value of $4^2 - 3 \times 2 - 6$?

A. 4

B. 16

C. 18

D. 28

2 A liter would be the most appropriate unit to measure which of the following?

A. the mass of a rock

B. the height of a tree

C. the amount of paper used to wrap a present

D. the amount of liquid in a jar

3 What is the value of $(0.2)(0.2)(0.2)(0.2)(0.2)$?

A. 0.00032

B. 0.0032

C. 0.0032

D. 0.32

4 The graph below represents the coordinates for which table of values?

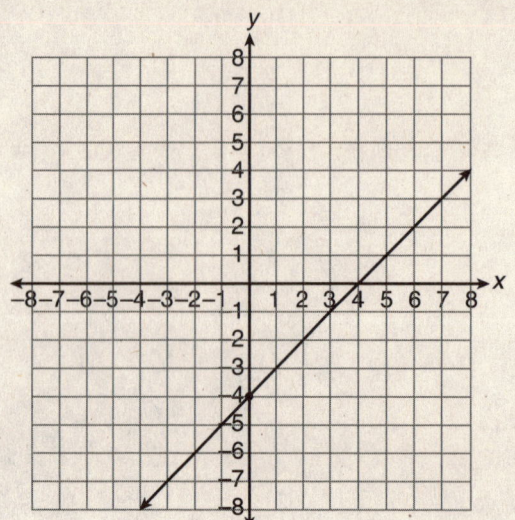

A.

x	y
−1	5
0	4
1	3

B.

x	y
−1	−5
0	−4
1	−3

C.

x	y
−1	3
0	4
1	5

D.

x	y
−1	−3
0	−4
1	−5

5 Which of the following mixed numbers is greater than $6\frac{1}{2}$?

A. $6\frac{1}{4}$

B. $6\frac{2}{3}$

C. $6\frac{2}{5}$

D. $6\frac{3}{8}$

6 The local newspaper held a contest asking local students to write an essay and/or create an art project that reflects school spirit. 58 students entered the contest. 30% of those students submitted art projects. Approximately how many students submitted art projects?

A. 5

B. 12

C. 18

D. 24

7 Tessa rolls a number cube labeled 1 to 6, then flips a coin. If the coin comes up heads, she multiplies the number she rolls by 4. If the coin comes up tails, she multiplies the number she rolls by 0. Which of the following is the set of possible outcomes?

A. 0

B. 1, 2, 3, 4, 5, 6, 4, 0

C. 4, 8, 12, 16, 24

D. 0, 4, 8, 12, 16, 20, 24

8 What is the value of $(-2)^5$?

```
┌─────────────────────────┐
│                         │
│                         │
│                         │
│                         │
└─────────────────────────┘
```

9 A bag of apples contains red apples, green apples, and yellow apples. If an apple is chosen at random, the probability that it is red is $\frac{2}{5}$. The probability that the apple is green is $\frac{1}{2}$, and the probability that it is yellow is $\frac{1}{10}$. If there are 12 red apples in the bag, what is the total number of apples in the bag?

```
┌─────────────────────────┐
│                         │
│                         │
│                         │
│                         │
└─────────────────────────┘
```

10 A coat is discounted at 10% off the regular price. The regular price is $65. How much is the discount?

```
┌─────────────────────────┐
│                         │
│                         │
│                         │
│                         │
└─────────────────────────┘
```

11 What is the next term in the following sequence?

1, 1, 2, 6, 24, 120, _____

A. 144

B. 600

C. 720

D. 1,200

12 Which of the following is closest to the square root of 65?

A. 8.1

B. 7.5

C. 6.2

D. 9.4

13 Evaluate the expression below for $m = -1$ and $n = -4$.

$-m(5 + n) - 2m$

A. −3

B. −2

C. 2

D. 3

14 Jose and Martin are playing a game. The 10 flashcards, numbered 0 through 9, are drawn randomly from a deck.

Jose will multiply the number on the flashcard drawn by 2. Martin will add 5 to the number on the card drawn. The player with the highest product or total at the end of the game wins.

a. Make a list of the possible outcomes for Jose when a card is drawn from the deck.

b. Make a list of the possible outcomes for Martin when a card is drawn from the deck.

c. What is the probability that Jose will have a score equal to or higher than Martin's when one flashcard is drawn? Explain your answer.

15 So far Kyle has walked about $\frac{2}{5}$ of the distance for a walkathon. Saira has just finished all 32 miles of the same walkathon. Approximately how much farther does Kyle have to go?

A. 10 miles

B. 12 miles

C. 18 miles

D. 25 miles

16 Brandon is driving to a town 600 miles away. The trip will take him a few days. By the end of the second day, he wants to have 150 miles or less left to drive. Which of the following inequalities shows the number of miles, m, Brandon will have to drive over each of the two days?

A. $600 - 2m \geq 150$

B. $600 - 2m \leq 150$

C. $600 - 150 \geq 2m$

D. $600 - 150 \leq 2m$

17 What is the length of the longest side of the right triangle shown in the diagram below?

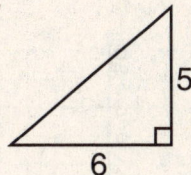

A. $\sqrt{25}$ units

B. $\sqrt{36}$ units

C. $\sqrt{61}$ units

D. $\sqrt{90}$ units

18 Triangle *XYZ* in the diagram below is rotated 90° clockwise about the point (–4, 3) and then reflected across the *x*-axis.

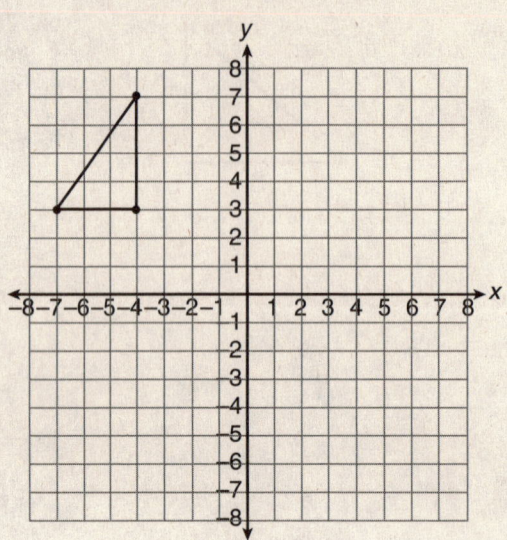

Which of the following is the image of △*XYZ* after the transformations?

A.

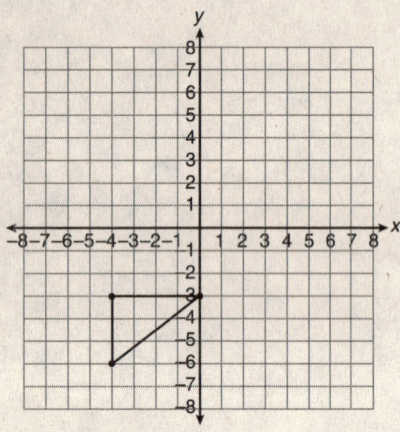

B.

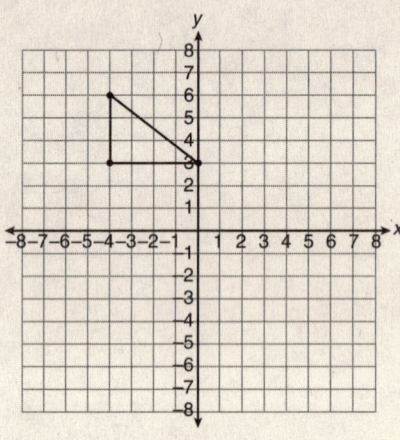

C.

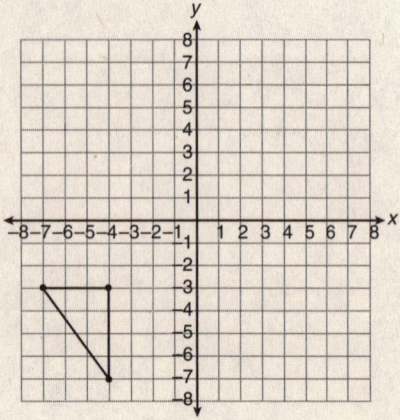

D.

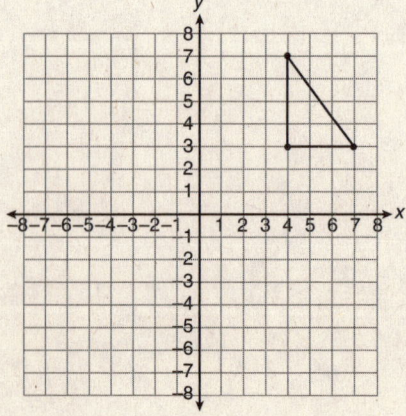

19 Briante asked 30 students about their favorite subject. 10 students said math was their favorite. Which portion of the graph below represents the students that chose math?

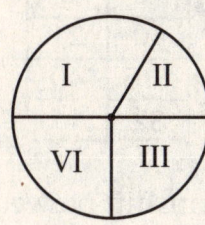

A. I

B. II

C. III

D. IV

20 Sonya wrote the numbers 1 through 10 on the board. If she selects one number at random, what is the probability that the number is evenly divisible by 3?

A. $\frac{1}{10}$

B. $\frac{3}{10}$

C. $\frac{2}{5}$

D. $\frac{1}{2}$

21 A group of students recorded the results of an experiment in the table below.

p	q
10	3
60	21
118	40
32	11

What is the approximate relationship between the two variables, *p* and *q*?

A. $p = \frac{1}{3}q$

B. $p = 3q$

C. $p = q + 7$

D. $p = q - 7$

22 A taxicab company charges a flat fee of $4.00 per ride and an additional $1.00 for each mile. What is the cost of 10-mile trip?

A. $4.00

B. $6.00

C. $10.00

D. $14.00

23 What is a possible ordered pair for the equation $3t + 4 = r$, if t is a negative integer greater than -4? Write the ordered pair in the form (t, r).

24 For a rectangle that measures 4 yards by 6 yards, express the area in square feet.

25 Ms. Woods is putting up a fence that measures 8 feet high and 22 feet across. What is the ratio of the width to the height?

26 Shaina is making a quilt using a pattern of equilateral triangles, as shown in the diagram below.

a. If the pattern continues, how many small equilateral triangles will be in the next (fourth) figure?

b. If the pattern continues, how many small equilateral triangles will be in the tenth figure?

c. Fill in the table showing the number of small equilateral triangles in the first ten figures.

Figure Number	Number of Triangles
1	
2	
3	
4	
5	
6	
7	
8	
9	
10	

d. What is the relationship between the figure number, *f*, and the number of small equilateral triangles, *t*?

27 A cube shown in the diagram below is folded from a single piece of paper. The pattern of paper that will fold into a cube is also shown below.

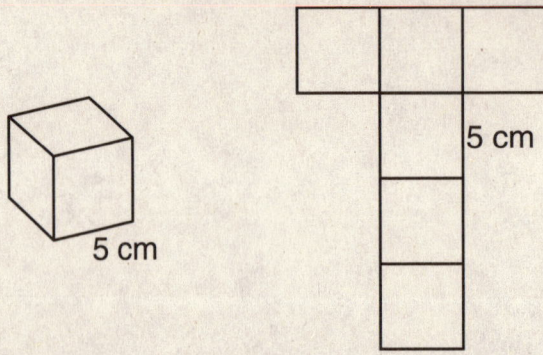

5 cm

5 cm

a. A cube has six congruent faces that make up its paper pattern. Draw a pattern made up of the same six congruent faces that will not fold into a cube.

b. Draw a paper pattern that will fold to make a rectangular prism with the same square base as above and twice the height. Label the lengths of your line segments.

28 Indra bought a car for $20,000. Each year, the value of the car depreciates. The chart below shows the value of the car after each year that Indra owns the car.

Years Owned	Car Value
0	$20,000
1	$17,500
2	$15,000
3	$12,500

According to the pattern in the chart, what will be the value of the car when Indra has owned it for 7 years?

A. $2,500

B. $3,000

C. $3,500

D. $5,000

29 Kipp is setting up booths for an art show. He needs to set up 11 booths. It takes him about 8 minutes to set up each booth. He wants to be finished setting up at 10:30 A.M. When should he start setting up the booths?

A. 8:30 A.M.

B. 9:00 A.M.

C. 9:30 A.M.

D. 10:00 A.M.

30 What is the value of $\frac{3}{4} \times \frac{1}{5}$ expressed as a decimal?

A. 0.15

B. 0.20

C. 0.5

D. 0.75

31 What is the product of x and its negative reciprocal?

A. -1

B. 1

C. $-x$

D. x

32 In a survey, 500 people were asked about the types of books they read for leisure. The circle graph below shows the results.

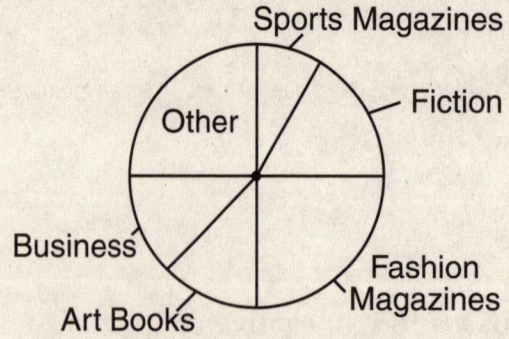

About how many people read fashion magazines for leisure?

A. 125

B. 200

C. 250

D. 300

33 Lola bought a bag of marbles. What percent of the marbles are black?

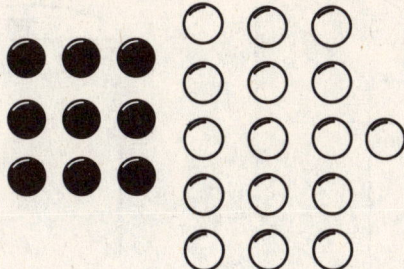

    A. 9%

    B. 36%

    C. 56%

    D. 64%

34 Which of the following has the same value as $\frac{x}{100}$?

    A. $0.01x$

    B. $\frac{1}{x}$

    C. $0.1x$

    D. $100x$

35 A rectangular prism and a cylinder are shown in the diagram below.

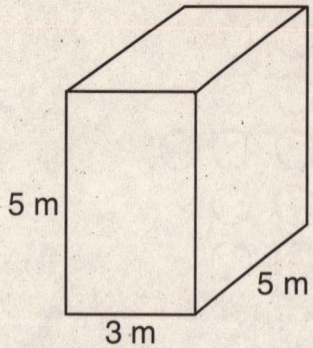 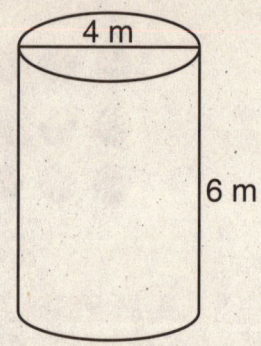

a. What is the volume of the rectangular prism? (Use your reference sheet to help you find the formula.)

b. What is the volume of the cylinder? Round your answer to the nearest whole number. (Use your reference sheet to help you find the formula.)

c. Based on your answer to parts *a* and *b*, how would the volume of the two solids compare if you doubled their dimensions? Justify your answer.

36 Brian made models of different rectangular prisms. He recorded the area of the base and the height of each prism.

Area of Base	Height	Volume
4	2	8
8	2	
16	2	32
4	4	16
4	8	

a. What is the volume of the rectangular prism when the area of its base is 4 and its height is 8?

b. Explain what happens to the volume of a rectangular prism when only the area of the base is doubled.

c. Explain what happens to the volume of a rectangular prism when only the height is doubled.

d. Based on your responses to parts *b* and *c*, explain what would happen to the volume if you doubled both the area of the base and the height of a rectangular prism.

37 What is w in the equation $14 = 14w - 2(5 + w)$?

A. 1

B. 2

C. 4

D. 8

38 In the equation $a = \frac{1}{b} - 3$, which of the following is true of a?

A. a must equal 0.

B. a must be a positive number.

C. a must be a negative number.

D. a can be either a positive or negative number.

39 Tasha had the following scores on her math tests:

78%, 92%, 78%, 57%, 97%, 82%, 83%.

What is Tasha's mean score?

A. 40%

B. 78%

C. 81%

D. 82%

40 Players are placed in the following age groups during a soccer tournament:

Age groups (in years)	Number of Players
3–5	26
6–8	38
9–11	32
12–14	44
15–17	24

One player will be selected at random to help with the award presentations at the end of the tournament. Estimate the probability that the randomly selected player will be younger than 12.

A. 2%

B. 44%

C. 58%

D. 96%

41 Which of the following values for x and y satisfy the equation $y = x^2 + x - 2$?

A. $x = 1, y = -2$

B. $x = 0, y = 2$

C. $x = 0, y = -2$

D. $x = 1, y = 2$

42 In the equation $xy = 3$, which is a true statement?

A. Either x or y is negative.

B. x must be a negative number.

C. y must be a negative number.

D. x and y must both be the same sign.

43 In the following diagram, *MN* and *OP* are parallel lines. Line segment *SB* bisects angle *NSR*. What is the measure of the angle marked *x*?

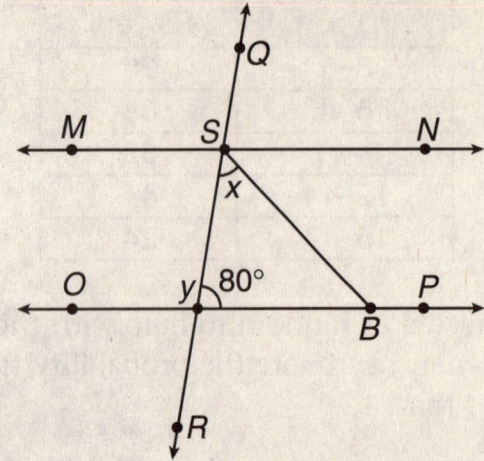

A. 10°

B. 50°

C. 80°

D. 90°

44 Li had $50 in his wallet. He bought 4 boxes of candles. Each box cost the same amount. He also bought a candleholder for $8. Which of the following inequalities shows the cost, *n*, of each box of candles?

A. $\frac{\$4}{n} \leq \$50 + \$8$

B. $\$4n + \$50 \geq 2n$

C. $\frac{\$n}{4} + \$8 \geq \$50$

D. $\$4n + \$8 \leq \$50$

45 About how many degrees are there between any two numbers on a clock?

A. 5°

B. 10°

C. 15°

D. 30°

46 The blades of the fan in the diagram below rotate so that one blade occupies the location of the previous blade.

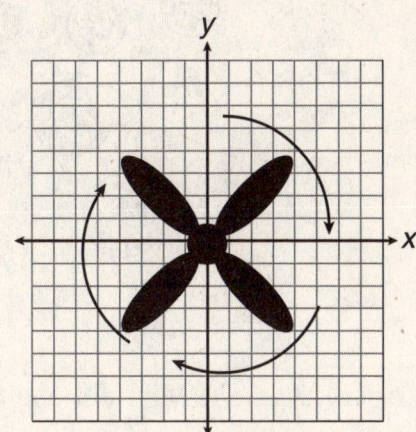

Which of the following describes the distance each blade moves to get to the next location?

A. 90° clockwise

B. 90° counterclockwise

C. 180° clockwise

D. 180° counterclockwise

47 A teacher ordered 7 large pizzas for a class party. Each pizza was cut into 8 slices. Each student in the class ate 3 slices. After the party, there were 5 slices left. Which equation can be used to find m, the number of students in the class?

A. $7m + 3 = 5 \times 7$

B. $7 + 3m + 8 \times 5$

C. $3m + 5 = 7 \times 8$

D. $3m = 7 \times 8 + 5$

48 Michael found the following coins in his pocket:

a. What combinations of coins could Michael use to deposit 25 cents in the snack machine?

b. Is it possible for Michael to select 50 cents using exactly 5 of his coins? Explain your answer.

c. What is the least amount of coins Michael can select to total 42 cents? Explain your answer.

Answers and Explanations for Practice Tests

ANSWERS AND EXPLANATIONS FOR PRACTICE TEST 1

1. **D** This problem asks you to find the best unit of measurement for the area of a sheet of notebook paper. Because you multiply two quantities, the length and the width, to find area, the units are square. Only choice **D** has units that are square, so you know that **D** is the correct answer choice.

2. **B** You need to use the order of operations to solve this problem. Remember that the order is Parentheses, Exponents, Multiplication and Division (in the order that they appear), and Addition and Subtraction (in the order that they appear). Because there are no parentheses, you evaluate the exponent 2^3 first to get 8. Then you evaluate the division, $(-12) \div 2$, to get -6. Finally, you do the addition and subtraction, $4 + (-6) - 8$, to get -10. So the correct choice is **B**.

3. **D** Estimate to solve this problem. To determine an amount that is about 10% more than 92, first round 92 down to 90. Then, find 10%, or 0.1, of 90. This is equal to 9. So 9 more than 92 is about 100. **D** is the correct answer choice.

4. **A** You need to determine which table of values corresponds to the graph. Each pair of x- and y-values in the correct answer choice will identify a point on the graph. Look at the values in the answer choices and see which points fall on the line. Because $(-1, 1)$, $(0, 3)$, and $(1, 5)$ are points on the line, the correct choice is **A**.

5. **B** This question asks you to find the number that has a square root of about 7.5. Remember that finding the square root is the inverse (or opposite) of squaring a number. You can solve this problem by squaring the given number: 7.5×7.5. Or you can estimate to find the answer. Because 7.5 is between 7 and 8, and $7 \times 7 = 49$ and $8 \times 8 = 64$, the answer is between 49 and 64. Only answer choice **B** is between 49 and 64.

6. **A** This question asks you to find the number that has a square root between 6 and 7. Remember that finding the square root is the inverse (or opposite) of squaring a number. Because $6 \times 6 = 36$ and $7 \times 7 = 49$, the answer is between 36 and 49. So the correct answer choice is **A**.

7. **D** When you flip a coin, the two possible outcomes are heads or tails, *H* or *T*. So the set of possible outcomes must include heads and a letter, and tails and a letter. You can eliminate answer choices **A** and **B** because those outcomes include either heads or tails, but not both. When you spin the spinner, you can get any one of the four sections of the spinner. So you can eliminate any answer choice that does not include all four sections. Answer choice **C** only includes sections *A, B,* and *C,* so you can eliminate this answer choice also. Answer choice **D** combines each letter with heads and then with tails. So **D** is the correct answer choice.

8. **B** To find the correct answer choice, you need to identify the number sequence which would have 46 as the sixth term. Look at the answer choices. They appear to increase with each consecutive number. You can eliminate answer choice **D** because the fourth term of that choice is 64 and 46 is less than 64. You can also eliminate answer choice **A** because the terms for that choice increase by 3 and the difference between the fourth term, 10, and the sixth term, 46, is so large that 46 could not possibly be sixth in the sequence. Answer choice **B** increases by multiples of 3, so 1 + 3 = 4, 4 + 6 = 10, 10 + 9 = 19, 19 + 12 = 31 (the fifth term), and 31 + 15 = 46 (the sixth term). So **B** is the correct choice.

9. **C** This question asks you to find an inequality for the scenario. Shannon bought 3 pairs of socks, so the amount she spent on socks can be represented as $3s$ (3 times the unknown cost of each pair). She also bought a shirt for $15, so she spent $3s$ + $15. This amount must be less than or equal to the amount of money she had, $25. So the correct choice is **C**. Make sure you look at the signs carefully before choosing an answer.

10. **A** Substituting the values into the expression would give you 5 + [−12 + 4(3 − 4)]. Follow the order of operations and you will get 5 + [−12 + 4(−1)]. This simplifies further to 5 + (−12 − 4), which is equal to −11. **A** is the correct choice.

11. 17 feet

This problem asks you to find the height of the tree reduced by 15%. First find 15% of the height, which is 0.15×20, or 3. Fiona trimmed 3 feet off the 20-foot tree. The new height of the tree is $20 - 3 = 17$ feet.

12. −36

Remember that -6^2 means $-(6 \times 6)$. The square is only on the 6 and not on the negative sign. So the product is −(36), or −36.

13. 4

The probabilities given are gray $\frac{4}{9}$, white $\frac{2}{9}$, and blue $\frac{1}{3}$. You also know that there are 3 blue socks, so 3 out of the total number of socks, x, should be equal to the probability $\frac{1}{3}$. Set up a proportion equating $\frac{1}{3}$ to $\frac{3}{x}$ and solve for x. By equivalent fractions, you know that $\frac{1}{3} = \frac{3}{9}$, so you know that there are 9 socks in total. Because $\frac{4}{9}$ is the probability of choosing a gray sock, you know that there are 4 gray socks out of the 9 socks in the drawer.

14. This problem has three parts.

a. You need to draw a pattern to form a triangular pyramid. Remember a triangular pyramid has a triangular base and 3 triangular sides. Each side of the base may be the same length.

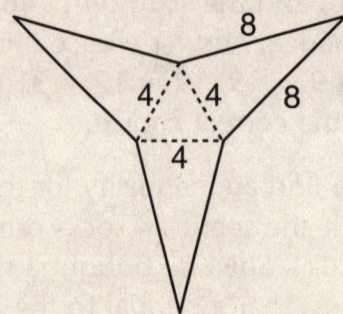

b. You need to draw a net that can be folded into a cone. Recall that a cone is formed from a triangular surface and has a circular base.

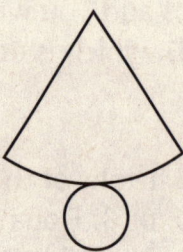

c. 37.68 units.

The circumference of the base can be found using the formula $C = 2r \times 3.14$. To find the circumference insert the radius and calculate: $2(6)(3.14) = 37.68$ units.

15. This problem has four parts.

a. 7

This part of the problem asks you to look at the pattern and determine the fourth term in the pattern. Because 1 musician was in the first song, 3 joined for the second, and 5 for the third, the pattern is that 2 more musicians join than the previous number that joined. Because 5 joined for the third song, 5 + 2, or 7, would have joined for the fourth song.

b. 25

From the pattern, you have 1 musician for the first song, add 3 for the second song, add 5 for the third song, add 7 for the fourth song, and add 9 for the fifth song, for a total of 25 musicians. The pattern also shows that the number of musicians is the square of the song number, so for song number 5 there would have been 5^2, or 25, musicians.

c. You have already found a pattern that shows the next odd integer joining each song or that the number of musicians equals the square of the song. Your table should look like the one below.

Song	Number of Musicians
1	1
2	4
3	9
4	16
5	25
6	36

d. From the table, you can see a pattern: $1^2 = 1$, $2^2 = 4$, $3^2 = 9$, $4^2 = 16$, $5^2 = 25$, and $6^2 = 36$. The pattern is that the square of the song number gives you the total number of musicians for that song.

16. **A** The percentage of computers at each grade level equals the number of computers divided by 200 then multiplied by 100. However, because a bar graph compares data, and you are really not asked to find the actual percentages, just compare the heights of the bars and find the two bars that add up to 80, the number of computers in the fifth grade. The two grade levels that combine for a total of 80 computers are grade 1 (30 computers) and grade 2 (50 computers). So choice **A** is correct.

17. **D** Find the hypotenuse of a right triangle by using the Pythagorean theorem, $a^2 + b^2 = c^2$, where a and b are the lengths of the legs (shorter sides) and c is the length of the hypotenuse (longest side). So $8^2 + 6^2 = c^2$, or $64 + 36 = 100 = c^2$. To solve for c you need to find the square root of 100, which is 10. So the correct choice is **D**.

18. **C** A reflection is a flip across a line of symmetry, in this case, the vertical axis. When a figure is reflected across the axis, each point of the image is the same distance away from the axis as the original point was. So the coordinates after the flip will be (4, 3), (4, 7), (7, 7), (7, 3). Because this image is then shifted 4 units to the left, each *x*-coordinate will now be 4 units less. The coordinates of the final image are (0, 3), (0, 7), (3, 7), (3, 3), or answer choice **C**.

19. **C** Because a monthly fee is charged in addition to the number of minutes, you must first calculate the charges for the minutes of usage each month, and then add the monthly fee. For the first month the cost for the minutes is $0.15 × 10, or $1.50. The cost of the minutes of usage plus the fee is $1.50 + $4.50, or $6.00. For the second month the cost for the minutes is $0.15 × 20, or $3.00. The cost of the minutes of usage plus the fee is $3.00 + $4.50, or $7.50. For the third month the cost for the minutes is $0.15 × 30, or $4.50. The cost of the minutes of usage plus the fee is $4.50 + $4.50, or $9.00. So the correct choice is **C**.

20. **B** You may know that 3 multiplied by itself four times is 81. Remember that when you multiply decimals, the number of decimal places in the product must be equal to the total number of decimal places in the factors. Each factor has one decimal place, so there are a total of four decimal places. The product must therefore have four decimal places. **B** is the answer choice with four decimal places.

21. **D** This problem asks you to find a probability. Recall that probability is the number of favorable outcomes out of the total possible outcomes. Because the favorable outcomes are NOT the vowels, they are the consonants in the word *SPELLING* and total 6 (SPLLNG). The possible outcomes are all the letters in the word *SPELLING*, which total 8. So the probability is $\frac{6}{8}$, which reduces to $\frac{3}{4}$. **D** is the correct choice.

22. **B** You are asked to change a fraction ratio to a percent ratio. To find the percent for 12 out of 25, or $\frac{12}{25}$, you can multiply the fraction by 100 to get $\frac{12}{25} \times \frac{100}{1} = 48$, 48%. Or you can set up a proportion to find an equivalent fraction out of 100, $\frac{12}{25} = \frac{x}{100}$. Solve for *x* by cross multiplication. $25x = 12 \times 100$, or $25x = 1,200$. So $x = 48$. The correct answer choice, 48%, is **B**.

23. **B**   Instead of finding the relationship between *s* and *g*, go directly to the answer choices and find the equation that best describes the relationship. Answer choice **A** says that *s* is about half as big as *g*. You know that all the values of *s* are bigger than the corresponding *g*, so **A** isn't right. The next two answer choices, **B** and **C**, seem to be reasonable. Answer choice **D** doesn't make sense because it also states that *s* is smaller than *g*. Now you are left to decide between **B** and **C**. Is *s* about two times *g* for all the values? Yes, but check choice **C**. Is *s* about 15 more than *g* for all the values? No, only for the last two values of *s*. So **B** is the best choice.

24. **D**   Because Nala has already traveled $\frac{1}{4}$ of the way, she has $1 - \frac{1}{4}$, or $\frac{3}{4}$, of the way left to go. Estimate the answer by rounding 206 to 200 miles and multiplying by $\frac{3}{4}$. Three-quarters of 200 is 150 miles, choice **D**.

25. 2:1

The ratio of the diameter to the radius is $\frac{d}{r}$. The diameter, *d*, of a circle is twice its radius, *r*, so $d = 2r$. Rewrite the equation to get the ratio $\frac{d}{r}$ on one side by dividing both sides of the equation by *r*. So $\frac{d}{r} = \frac{(2r)}{r}$, or $\frac{d}{r} = \frac{2}{1}$. The ratio of $\frac{d}{r}$ is 2:1.

26. (7, 17) or (9, 21)

You are told that in the equation $2a + 3 = b$, *a* is an odd number greater than 5 and less than 10. The only odd numbers greater than 5 and less than 10 are 7 and 9. So *a* can be 7 or 9. To find a coordinate pair in the form (*a*, *b*), you can substitute either 7 or 9 into the equation and solve for *b*. If you put $a = 7$ into the equation, then *b* is $2(7) + 3$, or 17. The ordered pair is (7, 17). If you put $a = 9$ into the equation, then *b* is $2(9) + 3$, or 21. The ordered pair is (9, 21). Either pair is correct.

27. $868.00

The cost of the sod is given in square feet, so you should calculate the area in square feet. There are 12 inches in a foot, so you should divide each measurement by 12 to find the dimensions in feet. The width, 84 inches, is equal to 84 in. ÷ 12 in., or 7 ft., and the length, 96 inches, is equal to 96 in. ÷ 12 in., or 8 ft. The area is 8 ft. × 7 ft. or 56 ft.^{2}. The cost to sod each square foot is $15.50, so the cost to sod 56 square feet is 56 × $15.50, or $868.00.

28. This question has three parts.

a. This part asks you to find the possible outcomes of the sum of the numbers on two cards. The following are the cards that could be selected and their sums:

    $1 + 2 = 3, 1 + 3 = 4, 1 + 5 = 6, 1 + 7 = 8, 1 + 9 = 10, 2 + 3 = 5, 2 + 5 = 7,$
    $2 + 7 = 9, 2 + 9 = 11, 3 + 5 = 8, 3 + 7 = 10, 3 + 9 = 12, 5 + 7 = 12,$
    $5 + 9 = 14, 7 + 9 = 16.$

    Arranging the sums in order gives you 3, 4, 5, 6, 7, 8, 8, 9, 10, 10, 11, 12, 12, 14, 16.

b. For player B to win, the sum on the cards must be less than 10. If 5 is the first card, then the remaining cards are 1, 2, 3, 7, and 9. To have a sum of less than 10, the cards 1, 2, or 3 must be chosen. So there are 3 favorable outcomes out of the remaining 5 cards. The probability is $\frac{3}{5}$.

c. The game is not fair. There are 8 ways to make a sum of less than 10, and only 7 ways to make a sum of 10 or greater. Because the probabilities are not equal, then the game is not fair.

Session 2

29. **D** This question asks you to divide a fraction by a fraction and then find the decimal equivalent of the quotient. To divide by a fraction, you change the divisor to its reciprocal and change the division sign to a multiplication sign. So $\frac{2}{5} \div \frac{1}{3} = \frac{2}{5} \times \frac{3}{1}$. Multiplying the numerators and the denominators gives you $\frac{6}{5}$. To change $\frac{6}{5}$ to a decimal number, divide 6 by 5 to get 1.2. So the correct choice is **D**.

30. **C** This question asks you to use the pattern in the table to determine the rate for 801–900 miles. For the first 100 miles, a customer pays $150, which is increased by $40 for each additional 100 miles. So from 100 miles up to 900 miles is an additional 800 miles. The additional charge for up to another 800 miles is $40 × 8, or $320. You must now add $320 for the additional miles to the $150 for the first 100 miles, for a total of $320 + $150, or $470. So the correct answer choice is **C**.

31. **B** This problem asks you to solve an inequality using an upper limit and a lower limit. Kris will work at least 20 hours, so he will earn at least 20 × $5, or $100. The most hours that he will work is 30, so he will earn no more than 30 × $5, or $150. Kris will earn between $100 and $150. Remember that a closed circle means that you include the point, and an open circle means that the point is not included. Because Kris could work 20 hours or 30 hours, both of those points are included. The number line that represents this scenario is **B**.

32. **D** You are asked to find the number of households that have fish for a pet using the Venn diagram. To do this, find the numbers within the circle labeled *Fish,* and add those numbers. Remember to add all the numbers in the *Fish* circle, even if they are in other circles as well. The sections of the Venn diagram that include fish are *Dogs and Fish* (3), *Cats and Fish* (4), *Dogs, Cats, and Fish* (6), and *Fish* only (28). The total for fish is 3 + 4 + 6 + 28 = 41. The correct answer choice is **D**.

33. **A** This problem asks you to find the answer choice that cannot be used for the equation. You cannot divide by 0, so *b* cannot be 0. Therefore, answer choice **A** cannot be true. *b* can be a negative number; that would simply make the fraction that includes *b* negative. So answer choice **B** is possible. *b* can be greater than *a*; for example, if *b* is a large number, then $\frac{1}{b}$ is a small fraction and *a* is negative. So answer choice **C** is possible. Both *b* and *a* can be negative; for example, if *b* is negative, the fraction $\frac{1}{b}$ is also negative, and subtracting 4 gives you a negative value for *a*. So answer choice **D** is also possible. The only answer choice that is not possible is **A**, so that is the correct choice.

34. **B** This question asks you to select the answer choice with the same value as $\frac{3}{x}$. Remember that $\frac{3}{x}$ means $3 \div x$. To divide, you can change the divisor to its reciprocal and change the division sign to a multiplication sign because division and multiplication are inverse operations. So $3 \div x$ is the same as $3 \times \frac{1}{x}$. Answer choice **B** is correct.

35. **A** This problem asks you to find an angle, given two parallel lines and a transversal. Recall the parallel line theorems. Angle *XPQ* is equal to angle *UQB* because they are corresponding angles of parallel lines. So angle *XPQ* is 75°. Angle *APY* is marked with the *m* and is the same measure as angle *XPQ* because they are vertical angles. So the angle marked *m* is 75°. The correct answer choice is **A**.

36. This question has three parts.

a. The combinations of two cards whose sum is 10 are as follows:

10, 0

9, 1

8, 2

7, 3

6, 4

b. The combinations of three cards whose sum is 10 are as follows:

0, 9, 1

0, 8, 2

0, 7, 3

0, 6, 4

1, 6, 3

1, 7, 2

1, 5, 4

2, 3, 5

c. The greatest number of cards that can be selected to get a sum of 10 is 5. The cards 0, 1, 2, 3, and 4 give you a sum of 10.

37. This question has three parts.

a. You need to find the volume of the cone. The formula for the volume of a cone is $\frac{1}{3}(3.14)r^2h$. The height, h, is 8 in., and because the diameter is 4 in., the radius, r, is $\frac{4}{2}$, or 2 in. Substituting these values into the equation gives you $\frac{1}{3}(3.14)(2^2) \times 8$. To the nearest hundredth, the volume of the cone is 33.49 cubic inches. Remember that volume is in cubic units.

b. You need to find the volume of the cylinder. The formula for the volume of a cylinder is $(3.14)r^2h$. The height, h, is 8 in., and because the diameter is 4 in., the radius, r, is $\frac{4}{2}$, or 2 in. Substituting these values into the equation gives you $(3.14)2^2 \times 8$. The volume of the cylinder is 100.48 cubic inches.

c. The formula for the volume of the cone is $\frac{1}{3}(3.14)r^2h$ and the formula for the volume of the cylinder is $(3.14)r^2h$. Because both shapes have the same height and radius, $3.14 \times r^2 \times h$ is the same for both the cone and the cylinder. The cone has to be multiplied by $\frac{1}{3}$ to get the volume, so the volume of the cone is $\frac{1}{3}$ the volume of the cylinder.

38. **A** You are asked to find the mode of the prices given. The mode is the number that occurs most frequently. Only $15.00 appears more than once. Because $15.00 appears most frequently, it is the mode, and the correct choice is **A**.

39. **A** The question asks you to find a probability in percent form. You know that a probability is a ratio of the favorable outcomes to the total possible outcomes. The favorable outcomes are the black hooded sweatshirts. There are 4 black hooded sweatshirts. The total possible outcomes are all the sweatshirts available, which is a total of $3 + 4 + 8 + 4 + 12 + 3$, or 34. $\frac{4}{34}$ as a percent is $\frac{4}{34} \times 100$, or 12% (rounded to the nearest percent). So the correct choice is **A**.

40. **A** This question asks you to find an inequality from information given in a word problem. Selena will have the carpet cleaner for 6 hours. The cost for 6 hours is 6 times the hourly cost. Because the hourly cost is unknown, the variable c is used. So the cost for 6 hours is $6 \times c$, or $6c$. You are told that the total cost for the 6 hours is not less than $30 and not more than $40, so $6c$ is greater than or equal $30, but less than or equal to $40. This translates into the mathematical inequality given in choice **A**.

41. **C** This question asks you to choose the correct equation for the given table of values. You will know that you have the correct equation when you substitute each x-value given and get the corresponding y-value given. For choice **A**, when $x = -2$ is substituted into the equation, the corresponding y-value is -5, as given. However, when $x = -1$ is substituted into the same equation, the corresponding y-value is -4, which is not given. So answer choice **A** is not correct. For choice **B**, when $x = -2$ is substituted into the equation, the corresponding y-value is 1, which is not given. So answer choice **B** is not correct. For choice **C**, when $x = -2$ is substituted into the equation, the corresponding y-value is -5, as given. When $x = -1$ is substituted into the same equation, the corresponding y-value is -3, again as given. When $x = 0$ is substituted into the equation, the corresponding y-value is -1, as given, and when $x = 1$ is substituted into the same equation, the corresponding y-value is 1, which is also given. So **C** is a correct answer choice. For choice **D**, when $x = -2$ is substituted into the equation, the corresponding y-value is -3, which is not given. So answer choice **D** is not correct, and answer choice **C** is the only correct choice.

42. **D** This problem asks you to solve for y in the equation. First you need to simplify the equation by using the distributive property to multiply $3(y + 5)$. Remember that you multiply the quantity outside the parentheses by all the quantities inside the parentheses when using the distributive property. So $3(y + 5) = 3y + 15$. Now the equation is $5y = 9 + 3y + 15$. You can add $9 + 15$ to get 24, which simplifies the equation even further to $5y = 24 + 3y$. Now you can get the variables, y, to one side of the equation by subtracting $3y$ on both sides of the equation. $5y - 3y = 24 + 3y - 3y$. This gives you the equation $2y = 24$. Finally, divide each side of the equation by 2 to get $y = 12$. The correct answer choice is **D**.

43. **C** This question asks you to identify the statement that cannot be true. a can be greater than 1 when b is negative and a fraction, for example if $a = 2$ and $b = -\frac{1}{8}$. This example helps you to eliminate answer choices **A** and **B**, because both are possible. For answer choice **C**, neither a nor b can be 0. If either is 0 then the entire equation equals 0. Because the equation is not equal to 0, then neither a nor b can be 0. So answer choice **C** is correct, because it is not possible. For answer choice **D**, either a or b has to be negative because the right side of the equation is negative and the only way to get a negative product is for one of the variables to have a negative value. So choice **D** is possible. Because only answer choice **C** is not possible, it is the correct choice.

44. **C** In order to make it to her recital, Anita must be there right before recital time. If she leaves at 6:30 she cannot be at the recital at 6:30. This eliminates answer choice **D**. Anita has to travel a distance of 35 miles, which is close to 40 miles. Because the train travels at 80 miles an hour, it goes 80 miles in one hour, and 40 miles in a half hour, or 30 minutes. So it will take her about 30 minutes to get to her recital. Thirty minutes before 6:30 P.M. is 6:00 P.M. **C** is the correct choice.

45. **A** The total amount that was spent plus the amount left over equals 75. Because you know the cost of each exhibit and the number of exhibits, you can use the expression 3 × 20 for the total cost of the exhibits. Because you do not know the cost of the souvenirs, you can use the variable s. So Montel's class spent 3 × 20 + s. They had $4 left over, so the total amount is now 3 × 20 + s + 4. This is equal to the amount the class raised, so 3 × 20 + s + 4 = 75. The correct equation is answer choice **A**.

46. **B** This question asks you to find an image point after a translation. You can read the location of point X in the original figure. This point, originally at (2, 5), moves to (–1, 0) after the translation. This means the point was translated 5 units down and 3 units to the left. Because point X undergoes this translation, point Z undergoes the same translation. The location of point Z in the original figure is (2, 2), so a translation 5 units down and 3 units to the left takes it to (–1, –3). The correct answer choice is **B**.

47. **B** This problem asks you to find a total number, given information about parts of the whole. You are told that $\frac{2}{5}$ of students had juice and $\frac{1}{3}$ had milk. So the total fraction for juice and milk is $\frac{2}{5} + \frac{1}{3} = \frac{6}{15} + \frac{5}{15}$, or $\frac{11}{15}$. $\frac{11}{15}$ of the students had juice or milk. The remaining students, $1 - \frac{11}{15}$, or $\frac{4}{15}$ had water. Because the remaining students total 24, $\frac{4}{15}$ of the total students = 24. So $\frac{4}{15}x = 24$, or $x = \frac{24 \times 15}{4}$, which is 90. There are 90 students altogether. Answer choice **B** is correct.

48. This question has four parts.

a. A hexagonal pyramid has a 6-sided base and 6 edges that form triangles, for a total of 12 edges.

b. An octagonal pyramid has 1 octagonal face that is its base and 8 triangular faces, for a total of 9 faces.

c. If you complete the table, you should find a pattern that shows you that the number of edges in any pyramid is twice the number of sides in the base. So the number of edges = 2 × number of sides in the base.

d. From the table, you should also find a pattern that shows you that the number of faces of any pyramid is 1 greater one greater than the number of sides in its base. So the number of faces = number of sides in the base + 1.

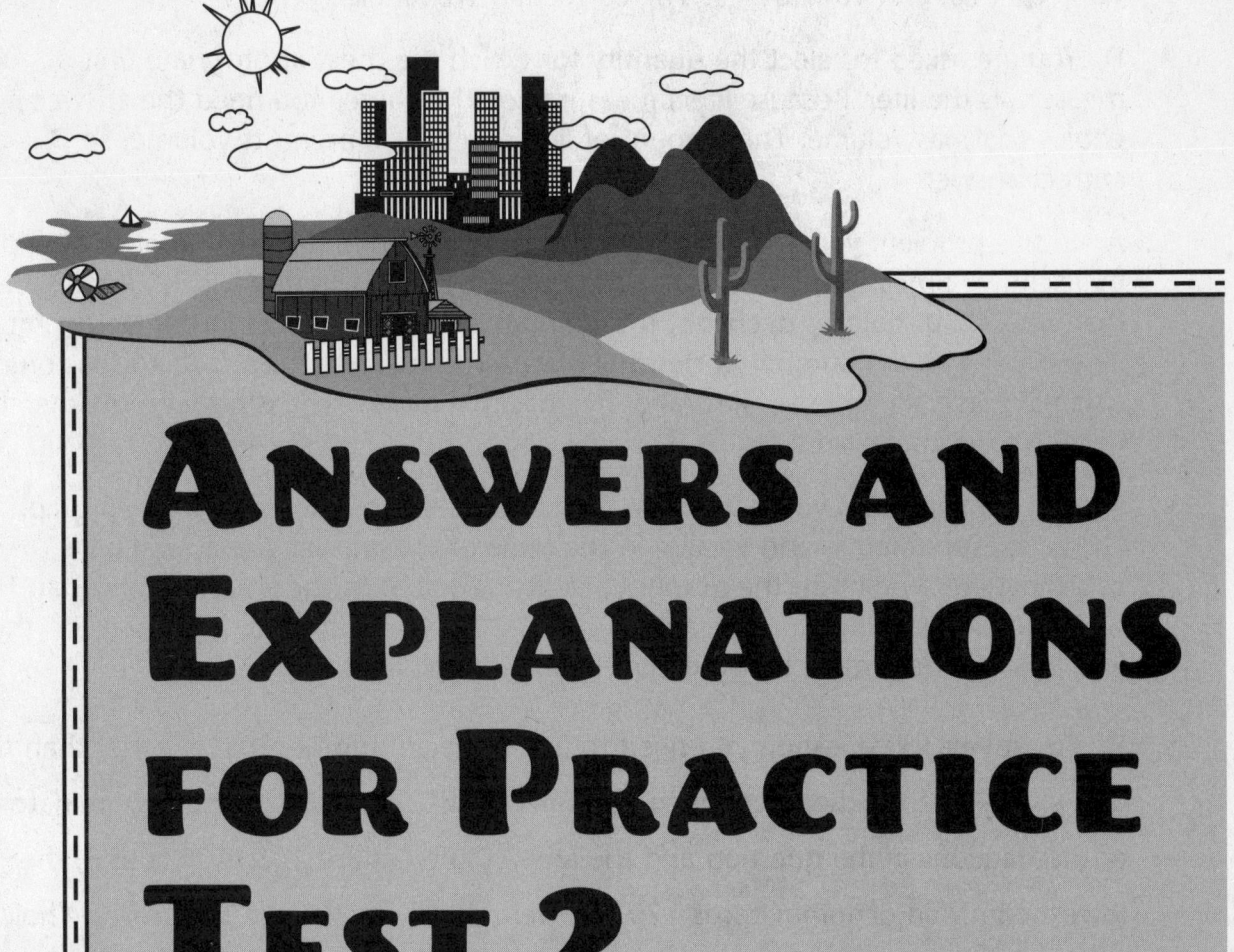

ANSWERS AND EXPLANATIONS FOR PRACTICE TEST 2

1. **A** You need to use the order of operations to solve this problem. Remember that the order is Parentheses, Exponents, Multiplication and Division (in the order in which they appear), and Addition and Subtraction (in the order in which they appear). There are no parentheses, so evaluate the exponent first: $4^2 = 16$. Next, evaluate the multiplication: $3 \times 2 = 6$. Finally, do the necessary subtraction: $16 - 6 = 10$, and $10 - 6 = 4$. The correct answer choice is **A**.

2. **D** You are asked to select the quantity for which the most appropriate unit of measure is the liter. Because liters measure liquid volume, you need the answer choice that has volume. The amount of liquid in a jar represents volume, so **D** is the correct answer.

3. **A** In this problem you are asked to find the product of a decimal number multiplied by itself five times. You know that 2 multiplied by itself five times is 32. Remember that when you multiply decimals, the number of decimal places in the factors must be the same as the number of decimal places in the product. Each factor has one decimal place, so there are a total of five decimal places. The product must therefore have five decimal places.

4. **B** This problem asks you determine the correct table of values for a given graph. Each corresponding x- and y-value in the table of the answer choice will be a coordinate of a point on the graph. Look at the values in the answer choices and see which ones fall on the line. Because $(-1, -5)$, $(0, -4)$, and $(1, -3)$ are points on the line, you can determine that the correct answer choice is **B**.

5. **B** To answer this question you need to find a mixed number that is larger than the one given. Change the fractions to equivalent fractions. Because the denominators of the fractions in the question and the answer choices are 2, 3, 4, 5, and 8, the lowest common denominator is 120. Convert all the fractions in the answer choices and find the fraction that is greater than $6\frac{1}{2}$, or $6\frac{60}{120}$. Answer choice **B**, $6\frac{2}{3}$, equals $6\frac{80}{120}$, which is greater than $6\frac{60}{120}$.

6. **C** There are a number of ways to solve this problem. Because the question asks for an approximate number, you can round off the number of students to 60 and find 30% of 60: $60 \times 0.30 = 18$. Choice **C**, 18, is the correct answer.

7. **D** Rolling the cube results in the following outcomes: 1, 2, 3, 4, 5, or 6. To find the effects of the coin landing on heads, multiply each of these outcomes by 4. The coin landing on tails means that each number cube result gets multiplied by 0. Because multiplying any outcome by 0 equals 0, add 0 to the list of possible outcomes. The outcomes are 0, 4, 8, 12, 16, 20, 24. **D** is the correct answer choice.

8. -32

   This question asks you to find the value of $(-2)^5$. Multiply (-2) by itself 5 times: $(-2) \times (-2) \times (-2) \times (-2) \times (-2) = -32$. Recall that multiplying two negatives gives you a positive, and multiplying a negative by a positive gives you a negative. The answer is -32.

9. 30

   The probability of selecting a red apple is $\frac{2}{5}$ and there are 12 red apples in the bag. The probability $\frac{2}{5}$ is equal to 12 red apples out of the total number of apples in the bag. Write a proportion: $\frac{2}{5} = \frac{12}{x}$. Solving for the unknown gives you 30. There are 30 apples in the bag.

10. $6.50

    This problem asks you to express a discount in dollars. The discount is 10%. Find 10% of 65 by multiplying $\frac{10}{100}$ by 65: $\frac{10}{100} \times 65 = 6.5$. The dollar amount of the discount is $6.50.

11. **C** Find the pattern of the number sequence. Each term of the sequence is multiplied by increasing integers: 1, 2, 3, 4, and so on, to the get the next term. The sixth term, 120, must be multiplied by 6 to get the next term: $120 \times 6 = 720$. The correct answer choice is **C**.

12. **A** Remember that finding a square root is the inverse (or opposite) operation of squaring a number. Ask yourself what number multiplied by itself gives you about 65. Because $8 \times 8 = 64$, the answer must be slightly more than 8. The only choice slightly larger than 8 is 8.1, which is answer choice **A**.

13. **D** Substituting the values into the expression would give you $-(-1)\,[5 + (-4)] - 2(-1)$. Follow the order of operations and evaluate the parentheses first: $[5 + (-4)] = 1$, giving you the expression $-(-1)(1) - 2(-1)$. Evaluate the multiplication next. This gives you the expression $1 - (-2)$. Now subtract the terms. Remember that subtracting a negative is the same as adding a positive: $1 - (-2) = 3$. **D** is the correct answer choice.

14. This question has three parts.

a. Jose will multiply the number on any card he draws by 2, so the possible outcomes or scores he can receive are

$2 \times 0 = 0$ $2 \times 5 = 10$

$2 \times 1 = 2$ $2 \times 6 = 12$

$2 \times 2 = 4$ $2 \times 7 = 14$

$2 \times 3 = 6$ $2 \times 8 = 16$

$2 \times 4 = 8$ $2 \times 9 = 18$

The possible outcomes are 0, 2, 4, 6, 8, 10, 12, 14, 16, and 18.

b. Martin will add 5 to the value of any card he draws, so the possible outcomes or scores he can receive are

$5 + 0 = 5$ $5 + 5 = 10$

$5 + 1 = 6$ $5 + 6 = 11$

$5 + 2 = 7$ $5 + 7 = 12$

$5 + 3 = 8$ $5 + 8 = 13$

$5 + 4 = 9$ $5 + 9 = 14$

The possible outcomes are 5, 6, 7, 8, 9, 10, 11, 12, 13, and 14.

c. $\frac{1}{2}$

Jose has a score equal to or greater than Martin's if the card selected is a 5, 6, 7, 8, or 9. There are 5 favorable outcomes out of a possible 10 outcomes. The probability of Jose having a score equal to or greater than Martin's is $\frac{5}{10}$, or $\frac{1}{2}$.

15. **C** Kyle has walked $\frac{2}{5}$ of the way, so he has $1 - \frac{2}{5}$, or $\frac{3}{5}$, of the way left to go. The entire distance of the walkathon is 32 miles. Round the distance to 30 miles and multiply by $\frac{3}{5}$: $\frac{3}{5} \times 30 = 18$. The correct answer is 18 miles, or **C**.

16. **B** The distance Brandon travels in 2 days can be represented by the number of days, 2, times the minimum number of miles driven each day, m, which is $2m$. Because the entire trip is 600 miles, the distance he has left is $600 - 2m$. The distance left to travel must not be greater than 150 miles, so the amount of miles he must have driven will be $600 - 2m \le 150$. The correct answer choice is **B**.

17. **C** To find a hypotenuse of a right triangle, use the Pythagorean theorem, $a^2 + b^2 = c^2$, where a and b are the legs and c is the hypotenuse. So $5^2 + 6^2 = c^2$, or $61 = c^2$. To solve for c you find the square root of 61. The correct answer choice is **C**.

18. **A** The triangle is rotated 90° clockwise about the point (–4, 3) and then reflected across the x-axis (horizontal axis). A rotation turns the original figure, and a 90° rotation is a $\frac{1}{4}$ turn. A $\frac{1}{4}$ turn clockwise moves the original figure to (–4, 6), (–4, 3), (0, 3). A reflection is a flip. When the image is flipped across the x-axis, the x-axis becomes the reflection line, or a line of symmetry. When reflected, the original figure moves to (–4, –6), (–4, –3), (0, –3). The correct answer choice is **A**.

19. **A** This question asks you to find a portion of a circle graph that represents 10 out of 30. 10 out of 30 can be expressed by the fraction $\frac{10}{30}$, or $\frac{1}{3}$. Section I represents $\frac{1}{3}$ of the circle. The correct answer choice is **A**.

20. **B** Probability is the ratio favorable outcomes to the possible outcomes. First, list all the favorable outcomes. The favorable outcomes are the numbers from 1 to 10 that are evenly divisible by 3: 3, 6, and 9. There are 3 favorable outcomes out of the total of 10 possible outcomes. The probability is $\frac{3}{10}$, answer choice **B**.

21. **B** This problem asks you to find an approximate relationship between the data in the table. One way to approach this question is to look at one column of data and then examine each corresponding value in the other column. The data in the column for p are about 3 times the amount of the data in the column for q. So $p = 3q$. The best answer choice is **B**.

22. **D** Calculate the cost of the mileage and then add the flat fee. The mileage is $1.00 per mile. Multiply the number of miles traveled by the cost per mile: $1.00 × 10 = $10.00. Add the $4.00 flat fee to the mileage total: $10.00 + $4.00 = $14.00. The correct answer choice is **D**.

23. (–3, –5), (–2, –2), or (–1, 1)

This problem asks you to find a possible ordered pair given an equation and a limit on the variables. The equation is $3t + 4 = r$, and t is a negative integer greater than –4. The only negative integers greater than –4 are –1, –2, and –3, so t can be –1, –2, or –3. Because you are asked for one coordinate in the form (t, r), substitute –1, –2, or –3 into the equation to solve for r. If $t = –1$, then r is 3(–1) + 4, or 1. The ordered pair is (–1, 1). If $t = –2$, then r is 3(–2) + 4, or –2. The ordered pair is (–2, –2). If $t = –3$, then r is 3(–3) + 4, or –5, and the ordered pair is (–3, –5). The possible ordered pairs are (–3, –5), (–2, –2), and (–1, 1).

24. 216 square feet

The rectangle measures 4 yards by 6 yards. Find the area of the floor in square feet by multiplying each dimension by 3. Expressed in feet, the rectangle measures 12 ft. by 18 ft., and the area is 12 ft. $\times$ 18 ft., or 216 ft.^{2}.

25. $\frac{22}{8}$ or $\frac{11}{4}$

The question asks you to find the ratio of the width of the fence to the height of the fence. The width of the fence is 22 feet and the height of the fence is 8 feet, so the ratio is 22 to 8, or $\frac{22}{8}$. The ratio can be simplified to $\frac{11}{4}$.

26. This problem has four parts.

a. 16

You can determine the pattern by looking at the number pattern or by looking at the geometric pattern. The first term has 1 small triangle, the second term has 4 triangles, and the third term has 9 triangles. The number of triangles in any term is the square of the term number. The fourth term must contain 4^2, or 16, triangles.

b. 100

The number of triangles in a term is the square of the term number. The tenth term will contain 10^2, or 100, triangles.

c. You have already found that the number of triangles is the square of the term number. So the completed table is as follows:

Figure Number	Number of Triangles
1	1
2	4
3	9
4	16
5	25
6	36
7	49
8	64
9	81
10	100

d. You can write the general equation as $f^2 = t$, if f is the figure number and t is the number of triangles.

27. This problem has two parts.

a. Draw the pattern with six squares in such a way that it cannot fold into a cube.

Some examples of patterns that cannot fold into a cube are shown below.

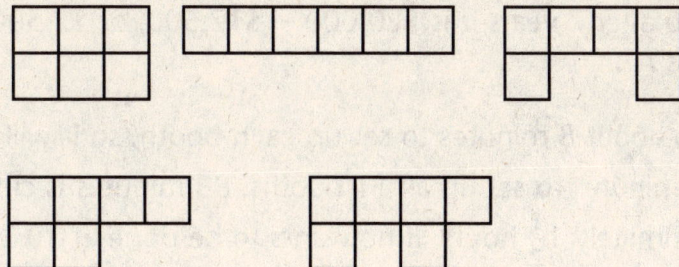

b. Draw a rectangular prism with a base of 5 cm by 5 cm and a height of 10 cm.

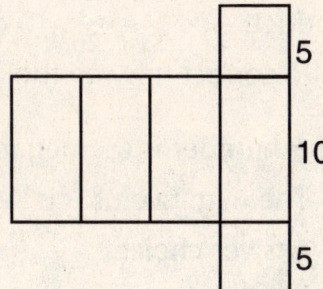

Session 2

28. **A** Use the pattern in the table to determine the value of the car after 7 years. According to the table, the value of the car depreciates $2,500 each year. After 7 years it depreciates $2,500 × 7, or $17,500. The value of the car at the beginning was $20,000, so after 7 years it is $20,000 − $17,500, or $2,500. The correct answer choice is **A**.

29. **B** It takes Kipp about 8 minutes to set up each booth, so it will take him about 11 × 8, or 88, minutes to set up all 11 booths. 88 minutes is close to 90 minutes, which is approximately $1\frac{1}{2}$ hours. If he wants to be done at 10:30 A.M., he will need to start $1\frac{1}{2}$ hours before that time, which is 9:00 A.M. **B** is the correct answer choice.

30. **A** This question asks you to multiply a fraction by a fraction and then find the decimal equivalent of the product: $\frac{3}{4} \times \frac{1}{5} = \frac{3}{20}$. To change $\frac{3}{20}$ to a decimal number, divide 3 by 20 to get 0.15. The correct answer choice is **A**.

31. **A** The negative reciprocal of a number is the negative of the number inverted. So the reciprocal of x, or $\frac{x}{1}$, is $\frac{1}{x}$. The negative of $\frac{1}{x}$ is $-\frac{1}{x}$. Find the product of x and $-\frac{1}{x}$: $x \cdot -\frac{1}{x} = -1$. **A** is the correct answer choice.

32. **A** The area occupied by fashion magazines appears to be about $\frac{1}{4}$ of the graph. Find $\frac{1}{4}$ of the total number of people surveyed: $\frac{1}{4} \times 500 = 125$. The correct answer choice is **A**.

33. **B** Nine out of the 25 marbles, or $\frac{9}{25}$, are black. Divide the numerator by the denominator to find $\frac{9}{25}$ as a percent: $\frac{9}{25} \times 100 = 36\%$. The correct answer choice is **B**.

34. **A** This problem asks you to find the answer choice that is equivalent to the given expression. $\frac{x}{100}$ is the same as $x \div 100$, or $0.01x$. **A** is the correct answer choice.

35. This question has three parts.

a. Find the volume of the rectangular prism. The formula for the volume of a rectangular prism is *lwh,* so the volume is $5 \times 3 \times 5$, or 75 m³. Remember that volume is always expressed is in cubic units.

b. Find the volume of the cylinder. The formula for the volume of a cylinder is $\pi r^2 h$, where π equals 3.14. The height, *h,* is 6 m, and because the diameter is 4 m, the radius, *r,* is $\frac{4}{2}$, or 2 m. Substitute these values into the equation: $(3.14)2^2 \times 6 = 75.36$ m³. Round the answer down to the nearest whole number, 75 m³.

c. The cylinder would have a slightly greater volume. The dimensions of the rectangular prism when doubled are 10 m by 6 m by 10 m, so the volume is $10 \times 6 \times 10 = 600$ m³. The dimensions of the cylinder when doubled are height = 12 m and diameter = 8, so the volume is $(3.14)(4^2)(12)$ or approximately 603 m³. Comparing the two, the cylinder has a slightly greater volume.

36. This question has four parts.

a. The formula for the volume of a rectangular prism is *lwh*. If the area of the base, or *lw,* is 4, then you multiply by the height, *h,* to get the volume. The volume is 4×8, or 32, cubic units.

b. Look at the formula for a rectangular prism again. It is *lwh*. If the area of the base is doubled, then you have 2*lw* and the volume is 2*lwh*. This is the same as 2(*lwh*), so the volume is doubled.

c. Look at the formula for a rectangular prism. It is *lwh*. If the height is doubled, then you have 2*h* and the volume is *lw* $\times$ 2*h*, which is the same as 2(*lwh*), so the volume is doubled.

d. Look at the formula for a rectangular prism once again. It is *lwh*. If the area of the base is doubled, then you have 2*lw,* and if the height is doubled, you have 2*h*. So the volume is 2*lw* $\times$ 2*h,* which is the same as $2 \times 2(lwh)$, or 4*lwh*. The volume is multiplied by 4.

37. **B** Solve for *w* in the equation. Simplify the equation by using the distributive property: $-2(5 + w) = -10 - 2w$. The equation becomes $14 = 14w - 10 - 2w$. Subtract $14w - 2w$ to get $12w$, which simplifies the equation even further to $14 = 12w - 10$. Adding 10 to both sides gives you $24 = 12w$. Finally, divide each side of the equation by 12 to get $2 = w$. The correct answer choice is **B**.

38. **D** Look at all the answer choices and test them out using the equation in the question. The choice that fits the equation is the answer. You can plug in multiple values for *a*, so answer choice **A** is not necessarily true. If you plug in 1 for *b*, $\frac{1}{1} - 3 = -2$, so you can see that *a* can be a negative number. Answer choice **B** is not necessarily true. If you plug in a very small number like $\frac{1}{5}$ for *b*, $1\frac{1}{5} - 3 = 2$, so you can see that *a* can be a positive number. Answer choice **C** is also not necessarily true. The variable *a* can either be positive or negative. The correct answer choice is **D**.

39. **C** Remember that the mean is the average. Calculate the mean by adding all the items and dividing by the number of items. The sum of $78 + 92 + 78 + 57 + 97 + 82 + 83$ is 567. Divide the sum by the number of terms: $567 \div 7 = 81$. **C** is the correct answer choice.

40. **C** The question asks you to find a probability in percent form. The favorable outcomes include all the players less than 12 years of age. Add together the numbers of players from the first three age groups (giving you all the players ages 3 to 11): $26 + 38 + 32 = 96$ players under age 12. The number of favorable outcomes is 96. The total number of outcomes includes all the players: $26 + 38 + 32 + 44 + 25 = 165$. $\frac{96}{165}$ expressed as a percent is $\frac{96}{165} \times 100$, or about 58%. The correct answer choice is **C**.

41. **C** If you substitute the given values from each answer choice, the one that makes the equation true is the correct answer. Substituting the values for choice **C** gives you $-2 = 0^2 + 0 - 2$. Evaluating the right side of the equation gives you -2, and the left side is -2, so the equation is true. These are the correct values, and the correct answer choice is **C**.

42. **D** Remember that multiplying two like signs (negative and negative, or positive and positive) gives you a positive product. Multiplying two unlike signs (a negative and a positive) gives you a negative product. The product of the two variables is a positive number 3, so the two values must have the same sign. **D** is the correct answer choice.

43. **B** $\angle QYP$ is congruent to $\angle QSN$ because they are corresponding angles. Because $\angle QYP$ is 80°, $\angle QSN$ is also 80°. You are told that line segment *SB* bisects $\angle NSR$, which means that it cuts it in half. $\angle QSN$ and $\angle NSR$ are supplementary angles, which means that they must add up to 180°. $\angle NSR$ is bisected, so it measures $2x$. Write an equation and solve for x: $80 + 2x = 180$. After subtracting 80 and dividing by 2, you should find that $x = 50°$, answer choice **B**.

44. **D** The total amount spent cannot be greater than the amount Li had. He bought 4 boxes of candles, each costing *n*, so the total cost of the candles is $4n$. He also bought a candleholder for $8. The total spent is $4n + \$8$, which must be less than or equal to $50. The correct answer choice is **D**.

45. **D** There are 12 numbers on the face of a clock. The clock is a circle, so there are a total of 360°. You can think of each number as dividing the clock into congruent pie pieces. Divide the total number of degrees in a clock by the number of pieces to find the degrees between each number: $360 \div 12 = 30$. There are 30° between each number. The correct answer choice is **D**.

46. **A** This question asks you to describe the turn of the blade. Because 4 blades are spread evenly throughout a "circle," the distance between each blade is $360° \div 4$, or 90°. The direction of turn is clockwise. The correct answer choice is **A**.

47. **C** Find an equation for the given scenario. There are 7 pizzas and each one is cut into 8 slices, so the total number of slices is 7×8. Each student ate 3 slices, with the number of students represented by the variable *m*. The total number of slices eaten equals $3m$. Because are 5 slices left, the number of slices eaten plus the number left over equals the total number of slices: $3m + 5 = 7 \times 8$. The correct answer choice is **C**.

48. This question has three parts.

a. List the combinations of coins that add up to 25 cents:

   1 quarter

   2 dimes and 1 nickel

   1 dime and 3 nickels

b. Yes. He could select 1 quarter, 1 dime, and 3 nickels to make 50 cents using 5 coins.

c. He can select 1 quarter, 1 dime, 1 nickel, and 2 pennies. This is the smallest combination of coins that will give Michael exactly 42 cents.